ANSWERS TO YOUR MOST CRUCIAL QUESTIONS ABOUT DHEA

—What is DHEA and why does the body produce it?

—Do men and women respond differently to DHEA?

—How does DHEA affect your mood, your looks, your sleep, your energy, your health, and your sex drive?

—Are over-the-counter DHEA supplements the same as natural DHEA?

—What dose of DHEA is safe and most effective?

— Can simple changes in your diet and lifestyle also help increase DHEA levels?

NOW FIND OUT EVERYTHING YOU NEED TO KNOW ABOUT ONE OF THE MOST TALKED-ABOUT MEDICAL DISCOVERIES OF OUR TIME!

DHEA

The Miracle Hormone That Can
Help You Boost Immunity,
Increase Energy, Lighten Your
Mood, Improve Your Sex Drive,
and Lengthen Your Lifespan

Maureen Callahan

A Lynn Sonberg Book

A SIGNET BOOK

SIGNET
Published by the Penguin Group
Penguin Books USA Inc., 375 Hudson Street,
New York, New York 10014, U.S.A.
Penguin Books Ltd, 27 Wrights Lane,
London W8 5TZ, England
Penguin Books Australia Ltd,
Ringwood, Victoria, Australia
Penguin Books Canada Ltd, 10 Alcorn Avenue,
Toronto, Ontario, Canada M4V 3B2
Penguin Books (N.Z.) Ltd, 182–190 Wairau Road,
Auckland 10, New Zealand

Penguin Books Ltd, Registered Offices:
Harmondsworth, Middlesex, England

First published by Signet, an imprint of Dutton Signet,
a division of Penguin Books USA Inc.

First Printing, February, 1997
10 9 8 7 6 5 4 3 2 1

Contents

Contents

Contents

Author's Note

This book is not intended to take the place of medical advice or treatment recommended by a trained medical doctor. All concerns regarding your health need to be directed to your physician, particularly if you have any preexisting illnesses or medical problems. Be sure to consult with him or her before following any of the suggestions or advice in this book.

In addition, keep in mind that the bulk of research using DHEA supplements has been conducted on laboratory animals. (Limited studies are available that look at DHEA use in humans, although researchers are beginning to slowly receive more funding for these endeavors.) All reasonable attempts have been made to include the most current and factual information and medical reports. However, there is no guarantee that future research findings, particularly from clinical trials done with people, might not change the recommendations made about this steroid hormone over time.

Introduction

Pills and potions touted as the "secret" to longevity or the next "fountain of youth" seem to have quackery written all over them. After all, how could one little pill turn off aging, rev up the immune system, boost libido, and do so many other miraculous things that keep a person young. Yet, outlandish as these promises may seem, experts are beginning to wonder if there might be at least some truth to the concept, particularly for a hormone with a lengthy and somewhat difficult to pronounce name—dehydroepidandrosterone (DHEA).

Produced by the adrenal glands, odd-shaped organs that sit atop the kidneys, levels of this hormone drop dramatically as people grow older, leading some scientists to speculate that this inexorable decline may be tied in some way to many of the negative outcomes of aging. As DHEA levels in the body begin to hit an all-time low, chronic illnesses such as heart disease, cancer, and adult onset diabetes begin to appear. Is it mere coincidence or are these two events intimately intertwined?

DHEA: What is all the fuss about?

Historically, hormones like DHEA (and last year's hot hormone melatonin) were not the first body chemicals looked to as a potential "fountain of youth." In the nineteenth century a French physiologist named Charles Eduoard Brown-Sequard injected himself with testicular extracts and professed that these substances made him young again. Of course, he really wasn't able to reverse the aging process. But because of his intuitiveness about matters chemical, Brown-Sequard is considered a founder of modern endocrinology.

Chances are if he were alive today, the concept that a hormone might slow aging would seem old hat. But in today's world, where more people are living longer, interest is sparked in almost any remedy purported to turn back the biological time clock. It's not always a pure medical interest, however. There's profit to be made with wrinkle creams, hair colorings to cover the gray, and now the latest antiaging elixirs, hormones. Companies are already wrangling to see who will hold the biggest share of the DHEA supplement market. Corporate pocketbooks are opening wide in the hopes that DHEA could be this year's melatonin. How do you separate the advertising hype about this hormone from what is solid scientific fact? How do you determine if there really is something of benefit to be had with DHEA supplements? It's not easy.

In fact, many stories in newspapers and magazines have oversimplified the research findings. Yes, there is a lot of legitimate research. For example, the National

Institutes of Health is supporting studies of DHEA. One is investigating the role of DHEA in Alzheimer's disease. The other is studying how DHEA influences the lifespan and course of disease in mice. But there are still many questions scientists would like to have answered about DHEA. Of course, companies that stand to make money off supplement sales and harried newspaper reporters on a tight deadline don't have the time, or in some cases the inclination, to give you the whole story. And forget asking a clerk in a health food store to bring you up to speed about DHEA. Chances are they'll sum up the hormone in two minutes or less like this: the body produces less DHEA as you grow older so therefore these lower levels are related to the chronic diseases of old age—heart disease, cancer, diabetes, etc. Secondly, studies indicate that treatment with DHEA can help relieve or prevent illnesses such as X, Y, and Z.

They neglect to tell you that most of the studies to date have been done with animals. Nowhere is there a mention of harmful side effects and risks of DHEA supplementation. And it's probably never mentioned what research they are using to back up these recommendations. In essence, the idea is to cash in on your interest in DHEA without filling you in on all the facts. And that's not the smart way to improve your health. Your best approach is to take the time to learn more about what DHEA may and may not be able to do for you. The aim of this book is to fill you in on all the facts and to point out the parts of the puzzle that are still missing. If the science is being "stretched" beyond what it can reasonably support that information will turn up, too. Because in the end, simply being an informed health care consumer could have a profound impact on how long you live and how well.

How to use this book

Medicine in the nineties has shifted far, far away from the "Marcus Welby" scenario where a physician instructs a patient, holds his or her hand, and the two work through an illness together. Managed health care and advances in technology have contributed to making the delivery of medical care more impersonal. In fact, much of the responsibility for keeping up with the latest medical issues now falls to the consumer. Not that there is anything wrong with this approach. But in today's medical setting, if you are that consumer, you'll need to be well informed in order to take an active role in your health care.

There is much excitement in the medical community about the potential health benefits of this potent adrenal hormone. But there is also a lot of hype and promotion that cloud the issue of who might benefit from DHEA, hype that is aimed at your pocketbook rather than your well-being. This book will address the issue of how DHEA might fit into your overall health care strategies and will help you sort through the myths, facts, and promises. It offers accurate details about how DHEA relates to a variety of chronic diseases, everything from cancer to diabetes to obesity. It is not meant as a textbook or medical review of every single study ever conducted on DHEA but rather as an overview of some of the more relevant findings.

Emphasis is on presenting a balanced picture, not simply the studies that support a role for DHEA but those that delve into the negative aspects of replacement hormone therapy as well. It's intention is to help you build a program for a longer, healthier life incorporating DHEA supplements or, if you choose, natural strategies aimed at boosting levels of this hormone. Toward that goal, the book is organized so that each chapter builds onto the next in a programlike fashion, giving you details about how DHEA can fit in with other prevention strategies for good health until you reach Chapters 11 and 12. At this point you will need to decide whether or not you plan to take DHEA supplements (under a physician's guidance) or whether you wish to implement lifestyle strategies to raise levels of the hormone naturally.

Of course, that's not to say that you can't skip around and read only the sections that interest you. For example, if heart disease runs in your family the discussion in Chapter 4 on DHEA and heart disease will be of primary interest. Yet don't let that stop you from reading other chapters. Good health is based on keeping all parts of the body healthy—not simply the heart. By the end of this book you should have enough knowledge to make an informed decision about where DHEA might fit in your life.

Chapter 1

DHEA: The Body's Master Hormone

It's easy to poke fun at the snake oil salesman of old. That his "cure-all" tonics and formulas sold well probably had as much to do with his marketing strategies as with a clientele that was hungry for potions and pills that might turn back the ravages of time or simply ameliorate illness. Fast forward to the 1990s and the medical climate is almost the same. There is a growing fascination with, and acceptance of, alternative medical treatments. Tired of high-cost health care and distrustful of the emphasis many doctors place on drugs and surgical treatments many Americans are looking for a kinder approach to health, one that harnesses the mind as a healer and looks to nature for cures, a perfect stage for the hormone DHEA.

That hormones such as DHEA are embraced by the public with so much enthusiasm only emphasizes the seriousness of this quest for natural healing. Banners in health food stores and even some pharmacies are labeling DHEA as an "antiaging drug" or "Fountain of Youth." Some of this is hype, yet there are clues that this all-important miracle substance might contribute to keeping you healthy, wealthy, and wise. To separate the real benefits from the hype you'll want to have a basic understanding of what hormones do in the body and where DHEA fits into the hormone family.

What is a hormone? What kind is DHEA?

The term hormone comes from a Greek word that translates to "arouse to activity" or excite. Originally biochemists felt these instigator-type substances were produced in one tissue (an endocrine gland) and sent via the bloodstream to act on another tissue (target organ.) Now, researchers realize that's too narrow of a definition. Here is what's important to understand about how hormones function:

• *A hormone may have one target tissue or it may act on several different tissues.* Steroid hormones (such as DHEA or sex hormones) have a similar chemical configuration yet each behaves in a different way; each targets different body tissues. For instance, progesterone, which is made in the ovaries, regulates a single function in a single tissue. It's target tissue is the lining of the uterus and it's job is to prepare the uterus for implantation of a fertilized egg. Another steroid hormone, aldosterone, also has a single function. But this function—to regulate sodium and potassium balance by directing sodium transport in and out of cells—is performed in several target tissues: the kidneys, the sweat and salivary glands, the gastrointestinal tract, and the parotid gland. Androgen hormones (such as testosterone) and estrogens conduct numerous functions, most of which relate to sexual reproduction, in a wide variety of different organs. DHEA, if research proves its potential, could also have many target tissues and a multitude of functions.

• *The process by which a target tissue responds to a hormone involves several steps.* In order for a hormone to act on a target tissue it either needs to gain entry to individual cells or find a way to work at the cell surface. Receptors on cells typically act as the key to the cell door. The hormone insulin is a perfect example of this hormonal entry process. Insulin is secreted by the pancreas to act as a carrier of glucose in the blood. What you might not realize is that insulin receptors on the cell's surface allow insulin to enter the cell. When these receptors are faulty or damaged, insulin and sugar can build up in the blood. (See Chapter 4 for more about this problem called insulin sensitivity.) In other words, receptors exert control over the influx of a hormone into and out of the cell. At the same time, hormones trigger other body chemicals to orchestrate changes in what is often a cascade effect; one change sets another in motion and another and another. In addition, key enzymes in the cell or body tissue may be necessary to help convert one hormone (DHEA, for example) into another (testosterone). Without those enzymes, conversion would not be possible.

• *The body employs a system of "feedback control" to keep hormones in precise balance.* The rate at which endocrine glands produce hormones is often influenced by other hormones. Scientists refer to the system as negative feedback control. That is, when one substance is present in abnormally high levels the body responds to this negative event by setting in motion a chain of reactions or a cascade that balances the system. Typically the master endocrine gland or pituitary, and the hypothalamus (the area of the brain that controls our sympathetic nervous system reactions such as the "fight or flight" response) orchestrate this feedback system.

Hormones: A delicate art of balance

Every minute of every day the body must constantly adjust to its external environment in order to function. Most of these minute adjustments are also influenced by negative feedback. For instance, if strenuous exercise or a too hot climate cause the body to overheat this signals you to start sweating in an effort to release excess heat and return the internal or core temperature to normal. Periodically throughout the day your heartbeat increases and decreases in response to stress. Scientists refer to all of these systematic attempts at balance as homeostasis.

Hormones are a big part of this "feedback control" homeostasis. These chemicals help the body to reestablish an internal equilibrium whenever any system is thrown "out of whack." It works something like the thermostat on your air conditioner or furnace. You set it at the desired temperature and when the internal thermometer within the thermostat senses an increase or decrease of a few degrees it immediately sends a signal to turn on the air conditioner or furnace to heat up or cool off the room. If you are attempting to cool the room, the cooler air eventually lowers the temperature and that lower temperature registers on the thermometer. The air conditioner automatically shuts itself off.

The only difference with the body is that the control mechanism is not a thermometer but nerve impulses and changes in the levels of chemicals in the body. Perceiving these signals, the body's various endocrine glands release enough of certain hormones to bring the body back into balance. In other words, the mechanisms needed for intercellular communication, the mechanisms that let us respond and adjust to a constantly changing external and internal environment involve two body systems, the nervous system and the endocrine system. Scientists view the nervous system as more or less of a fixed structure through which messages are conducted. The endocrine system is more of a mobile message unit; various hormones secreted by specific glands are transported throughout the body to act on adjacent or distant tissues.

An overview of the endocrine system

Experts say that one of the most remarkable features of the endocrine system is that it gives the body a number of different ways to solve a problem. Here are some of the key players and their primary roles.

The Endocrine System

Gland	Main Hormones Produced	Primary Functions
Adrenals	DHEA	Precursor to androgens. New roles being explored.
	cortisol, epinephrine, norepinephrine	Reaction to stress, "fight or flight"
	aldosterone	Maintenance of sodium & potassium levels

Gland	Main Hormones Produced	Primary Functions
Ovaries	estrogen, progesterone	Female sexual characteristics, pregnancy, menstruation
Pituitary	ACTH	Control other glands
	human growth hormone	Body growth
	FSH and LH	Control sex glands
	TSH	Control thyroid
Parathyroid	PTH	Control calcium & phosphate levels
Pancreas	insulin, glucagon	Control blood sugar
Pineal	melatonin	Control time clock
	serotonin	Brain function, mood
Testes	testosterone	Male secondary sexual characteristics
Thymus	thymosine	Control immune system
Thyroid	thyroxine	Metabolism
	calcitonin	Absorption of calcium, bone formation

DHEA: The "Mother Steroid"

With this background on the endocrine system, you can see that DHEA is but one of many hormones involved in the control of body functions. (In fact, the chart above only covers some of the major hormones; there are many more hormones and hormone derivatives with all kinds of minor functions.) Yet researchers William Regelson, Roger Loria, and Mohammed Kalimi of the Medical College of Virginia (at Virginia Commonwealth University) make a case for DHEA as what they call the "Mother Steroid." In a paper published in the *Annals of the New York Academy of Sciences* these scientists point out that DHEA may be more than just a precursor agent that can be converted, as needed, into testosterone or estrogen. It could have numerous functions of its own: controlling insulin levels, enhancing memory, blocking tumor formation, acting as a "buffer" against overproduction of hormones involved in stress-induced injury.

Much of this work is preliminary, but since DHEA was discovered back in 1934, there is a lot that scientists do understand about the chemical profile and nature of this hormone. First, it is produced by the adrenal glands, endocrine glands that sit atop the kidneys. Peak production appears to be in the early morning hours. Typically the body links unbound DHEA with a compound called sulfate to allow it to travel in the blood for long periods. DHEA alone has a short lifespan; the new compound, DHEA sulfate (DHEAS), is more stable. In fact, if you take an oral DHEA supplement chances are most of it will be rapidly converted to the more stable form, DHEA sulfate. *Note:* You'll see DHEA sulfate referred to in different ways in the book such as DHEA sulfate, DHEA-S, DHEAS because different scientists list this form of the hormone using different monikers. Each time a new report is covered the moniker used by that scientist is listed.

How hormones influence the "fight or flight" response

When scientists suggest a possible link among DHEA, stress, and another adrenal hormone called cortisol, an easy way to understand the connection is to look at what happens when the body responds to danger. If you've ever found yourself stranded on a dark, deserted highway or face to face with a vicious stray dog chances are that fearful experience triggered a strong physiological response. Whenever the body senses a threat of any kind that arouses the sympathetic branch of the autonomic nervous system. Heart rate increases, blood flow to the muscles increases, pupils dilate, and the hair literally stands on end—all changes that prepare you to flee the danger or put up a good fight.

What you probably didn't realize is that hormones released from your adrenal glands (epinephrine, norepinephrine, dopamine) helped to orchestrate that response. They don't expedite the "fight or flight" response alone but are helped by other hormones including glucocorticoids produced by the adrenals such as cortisol. This same hormonal chain of events can also be triggered in cases of anxiety (stress) or with panic- and stress-related disorders. In fact, the longterm effects of physiological and psychological trauma on endocrine function are something scientists continue to study.

Questions that still need answering

After considering the important roles that hormones play in various body functions, it's very easy to put forth an argument that is "pro" DHEA replacement therapy. After all, why would the body flood the system with such large quantities of a chemical that doesn't do anything. (Estimates are that at its youthful peak the body produces about 25–50 milligrams per day of DHEA and DHEA sulfate combined.) Consider, too, the overwhelming amount of animal studies that seem to indicate this hormone is a promising therapeutic candidate for everything from lupus to diabetes to weight control. Yet, before you get too caught up in the hoopla, realize that many scientists feel there are still some issues that need clarifying. Before Americans throw caution to the wind and load up on DHEA supplements many experts would like to be able to answer these questions:

• Why are there such profound differences in DHEA response between men and women in many of the preliminary research studies? Could it be possible that DHEA might be a hormone that is beneficial to men but not to women?

• Is it truly DHEA that is having an impact on all these body systems or could it be that some of the less well-known steroid breakdown products (metabolites) of DHEA are exerting an effect? Similarly, does DHEA act as an "active" agent or are some of the positive changes really the result of this precursor hormone's conversion into estrogen and testosterone?

• And perhaps most important, what is the underlying mechanism by which DHEA works. Until this mechanism is outlined the use of DHEA (even in medical studies) is more of a hit-and-miss therapy.

How solid is the science?

The fact that there are still many unanswered questions hasn't dampened the enthusiasm of researchers. Early findings yield some tantalizing clues about how DHEA may enhance immune function; that one change alone could influence many of the chronic diseases of old age. And clinical studies using DHEA to treat the autoimmune disorder lupus are well into gear and look very promising. It's just not something researchers have ever discussed with the general public before. As is the case with many scientific undertakings, scientists researching DHEA have labored on in relative obscurity for several decades.

Of course, that all changed back in 1995 when two major scientific meetings—one in the United States and one in Canada—brought this potent hormone into the spotlight. At the Canadian meeting researchers narrowed their focus to DHEA intracrinology. That is, the bioconversion of DHEA in tissues or cells into testosterone and estrogen. Of course media coverage was scant with all the chemical and cellular jargon that seemed to have little practical application. But the mere title of the proceedings of another conference in Washington, D.C., sponsored by the New York Academy of Sciences, "DHEA and Aging" invited a media avalanche. Reporters, like the reading public, are eager for any scientific findings that hint at the prospect of a substance that might help people stay young.

As top researchers in the field gathered to report their findings to colleagues, media and corporate America were checking out the next potential miracle antiaging elixir. Those findings, as well as some newer research reports, form a convincing body of evidence that DHEA is a potent hormone with amazing health potential. Keep reading and you'll find out why.

Chapter 2

Staying Young, Aging Well

If there was a way to reset your internal biological time clock, not just physically turn it back like the odometer on a car but actually change it so that your body feels like it did when you were in your twenties or early thirties, would you be interested? Of course. It's human nature to want to improve the quality of life, to feel the best that you can. And while they secretly cringe at the label "Fountain of Youth" that is being given to the hormone DHEA, scientists can't ignore some of the amazing findings that seem to suggest that this powerful internal body chemical, manufactured by the adrenal glands, might have a hand in turning back the aging clock.

DHEA is really not new. Researchers have known about and studied this androgen hormone for decades. But DHEA gained even more prominence, and a chance at the media spotlight when the New York Academy of Sciences hosted an international scientific meeting on "Dehydroepiandrosterone (DHEA) and Aging" in 1995. Close to forty different research groups presented papers on studies that looked at DHEA's involvement and connection to everything from heart disease to immunity to weight control.

Most of us are mainly concerned with the practical side of the research. Can this hormone erase the devastation that comes with the march of time, not just wrinkles and gray hair but physical changes that signal the onset of chronic degenerative diseases such as arthritis, Alzheimer's, and heart disease? Perhaps. This is the million-dollar question that continues to occupy scientific researchers. It all started when researchers noticed a link between levels of DHEA in the blood and age.

Supplies dwindle with age

The fact that levels of the adrenal steroids DHEA and DHEA sulfate decline markedly as people age, and in a decidedly progressive downward slope and fashion, is one of the prime reasons scientists are so intrigued. (By comparison, the adrenal hormone cortisol remains at normal or near-normal levels throughout life.) Experts contend that this gradual decline raises the distinct possibility of DHEA(S) being one of the cogwheels of the hypothetical biological clock, which determines lifespan.

Most of the studies that measure DHEA levels, however, do so in a cross-sectional segment of the population. Few studies have actually followed individuals and measured their DHEA levels throughout the lifespan. Nevertheless, cross-sectional studies show that for most people DHEA levels begin a gradual descent sometime in the second or third decade. By the age of sixty you will probably have about 15 to 20 percent of the DHEA that you had at twenty-five or thirty. In other words, the older you are, the less DHEA you are likely to have.

In 1994, a group of scientists from Quebec measured the serum concentrations of twenty-six steroids in a group of more than 2,400 men aged forty to eighty. They noticed that the drop in DHEA was dramatic compared to all of the other steroid hormones. Researchers even have data on DHEA levels in the oldest old. A population-based gerontological study in Leiden, The Netherlands, determined DHEA levels in 138 elderly men and women (all over the age of eighty-five) and compared them with a control group of sixty-four men and women aged twenty to forty years. In both sexes, DHEAS levels were approximately four times higher in younger adults than in the oldest old.

Studies like these will help scientists establish reference values for DHEAS levels not only in the healthy oldest old but across the lifespan, values that may be helpful down the road when researchers are ready to establish treatment regimens based on DHEA replacement therapy.

What does this downward shift in DHEA mean?

First, however, scientists must figure out what this downward shift in DHEA really means. It could be that DHEA is simply a biomarker (biological change) that signals aging but has no impact on the changes wrought by aging. Or there could be a direct correlation between the aging and DHEA. As DHEA levels drop there is a noticeable rise in the development of chronic degenerative diseases such as heart disease and cancer. It follows that if the two conditions are related, then correcting lower levels of DHEA might help turn off the processes that cause aging or in essence help recapture youthful vigor.

It's a logical assumption, albeit a simplified one. Nevertheless, a study done in the early eighties confirms that it can and does happen with animals. When scientists at Temple University administered DHEA to laboratory mice they noticed older animals took on not only a youthful appearance but also a youthful physiological state: pelts became sleek and glossy, obese bodies shed excess fat, cancers disappeared. It wasn't until 1994 and 1995 that researchers were able to test this "return to youthful vigor" concept with human subjects. Not surprisingly, the results were quite positive.

DHEA supplements: Some of the first clinical trials with people

Some of the most talked about clinical trials with supplemental DHEA, conducted by noted DHEA researcher Dr. Samuel Yen and his colleagues at the University of California at San Diego, suggest that people respond in a similarly positive way to DHEA hormone replacement as animals. Yen and his research group administered 50 and 100 milligram doses of DHEA to a group of middle-aged men and women (ages forty and up) in two different double-blind, placebo-controlled clinical trials. One trial lasted six months, the other was twelve months in duration.

Researchers proposed that restoring extracellular levels of DHEA and DHEAS in the volunteers to levels commensurate to those in young adults might yield beneficial health effects. In one six-month program, within two weeks after starting DHEA, serum levels of DHEA and DHEAS of participants were restored to those found in young adults and remained at those levels throughout the three-month treatment period. While there was no significant change in libido, insulin sensitivity, body fat, or lipid levels, most participants (84 percent of women and 67 percent of men) did report increased feelings of physical and psychological well-being. On the list of reported improvements: increased energy, improved mood, better quality of sleep, better ability to deal with stress.

(*Of particular note to women:* concentrations of androgen hormones, masculinizing chemicals responsible for secondary male sex characteristics, doubled during the treatment phase in women volunteers. However, researchers say they were within the range found normally in younger women.)

After promising results in the first trial (1994), researchers decided to double the dose of DHEA and extend the duration of replacement therapy to see what kind of impact that might have on some of the same biological endpoints. Fewer participants (eight men and eight women) and an older age group (fifty to sixty-five years was the range) make the results less generalizable.

Again blood levels of DHEA increased (1995). Lean body mass (muscle tissue) increased in both men and women. Body fat levels decreased in men but not in women. Muscle strength, measured by knee extension/flexion, increased in men but not in women. For some unknown reason, the researchers did not report on whether or not this group of volunteers experienced increased energy or improvements in psychological well-being with the greater dose.

Again androgen hormone levels tripled and quadrupled on the larger dose of DHEA and were well above what's considered the normal range for adult women. One female volunteer developed facial hair, although the problem resolved itself after the study ended. Researchers suggest that 100 milligram doses of DHEA for durations as long as six months may be excessive for women in light of potential changes in androgens (masculinizing hormones). Concern is that these levels could increase over time if supplementation were a permanent therapy. The scientists caution that more research is needed to determine how gender differences influence response to DHEA therapy.

What does it all mean?

Not to put a damper on this research, but no matter how exciting the findings, two small studies are not enough on which to base large-scale clinical recommendations and most DHEA experts are pushing for bigger clinical trials. French researcher Etienne-Emile Balieu, another major presence in this specialized field of research, is currently studying DHEA levels and the use of DHEA supplements in a portion of the population in his home country. (One of the people he is studying is Jeanne Calment, a Frenchwoman who turned 121 in February 1996 and is the oldest living person whose age has been confirmed.) Researchers at Baylor College of Medicine have conducted several clinical trials with postmenopausal women and another trial with men and women is underway.

The genetics and biology of lifespan

DHEA aside for a moment, what makes one person live to the age of 100 and another die of a heart attack at age fifty-five? That's the question gerontological researchers would love to be able to answer. At this point the longest documented lifespan in the U.S. is 121.5 years. Of course, not everyone lives this long since most of us fall victim to chronic illnesses such as heart disease and cancer much earlier in life. However, one research report estimates that if all the major causes of death after age fifty were eliminated, life expectancy could increase to ninety-five years.

Some scientists suggest food deprivation may be the key to longevity. If the degree that mortality rate slows in laboratory animals on calorie-restricted diets (with food supply virtually cut in half) could be applied to people, human life expectancy could approach 120 years. Crunching numbers and using elaborate calculations some scientists at Duke University say it's even plausible that someone might live as long as 130 years.

Yet, how long you live is as much contingent on your personal genetic blueprint as on environmental factors including calorie restriction or DHEA levels. Indeed, researchers studying some of the oldest old wonder if genetics might explain why some people are particularly resistant to chronic degenerative illnesses that disable or kill most people before the ninth decade. Is some kind of "survival-of-the-fittest phenomenon" happening? wonders one Harvard geriatrician, a primary investigator on the New England Centenarian Study. Present knowledge holds that many interrelated factors including genes (there may even be a group of "longevity" genes) and perhaps lifestyle play a role in determining our bilogical time clocks.

Some lessons in longevity

Granted, genetics plays a big role in determining how long any person lives. But the extent of that role is not carved in stone. It's the old dilemma of nature versus nurture. How much of why we age is due to environmental factors and how much is programmed by a master genetic blueprint? Hoping to unravel the mysteries of why some people live longer than others, a group of California scientists assumed the herculean task of following the health habits of several thousand of the residents of Alameda county. The study is more than a decade old but what these researchers discovered was a definite link between certain health habits and behaviors and lifespan. Their landmark findings indicate that people who lived the longest seemed to have these seven practices or strategies in common:

1. They exercised regularly.
2. They kept their weight at a desirable level for height.
3. They routinely ate breakfast.
4. They didn't snack between meals.
5. They slept seven or eight hours each night.
6. They drank alcohol only moderately or not at all.
7. They had never smoked a cigarette.

Updating longevity advice for the nineties

While the Alameda "longevity-promoting" strategies are still valid, they don't incorporate the changing face of medicine and health research in the nineties. Since the Alameda report scientists have discovered antioxidant compounds that may help slow aging, genes that may predict who succumbs to illness and who doesn't, and, of course, hormones such as DHEA that may help turn back the aging clock. Recently the National Institute on Aging, a branch of the National Institutes of Health, devoted an entire "Age Page" report to the issue of Life Extension and current strategies that are being promoted to slow aging. The experts took a cautious stand on DHEA supplements, stating that no one knows if they are effective. Here's a brief synopsis of what these experts have to say about other new anti-aging strategies you might be thinking about employing:

• **Antioxidants**—It is well known that these natural substances are adept at fighting harmful molecules called free radicals, unstable compounds which can wreak havoc on body tissues. Normally, the body's internal antioxidant defense system prevents most free radical damage. But experts theorize that as people grow older defense systems may become overwhelmed by increasing amounts of free radical damage. Left unchecked that damage may eventually cause cells, tissues, and organs to break down leading to chronic illnesses of old age such as heart disease, cataracts and certain types of cancer.

Some antioxidants, such as the enzyme SOD (superoxidase dimustase), are produced in the body. Others come from food; these include vitamin C, vitamin E, and beta-carotene, which is related to vitamin A.

Most experts think that the best way to get these vitamins is by eating fruits and vegetables (five or more helpings a day) rather than by taking vitamin pills. SOD pills have no effect on the body. They are broken up into different substances during digestion. More research is needed before specific recommendations can be made.

• **DNA and RNA**—DNA (deoxyribonucleic acid) is the material in every cell that holds your genetic blueprint and therefore acts somewhat like an instruction handbook to the cell telling it how to behave and respond. Every day some DNA is damaged and must be repaired. But as more and more damage occurs with age, it may be that DNA repair, never 100 percent perfect, falls further and further behind. If so, the damage that does not get repaired could be one of the reasons that people age.

Unfortunately, over-the-counter tablets containing DNA and RNA (ribonucleic acid, which works with DNA in the cells to make proteins) cannot reverse this damage nor can they stand in for your own DNA. When taken by mouth, these tablets are broken down into other substances and cannot get into your body cells or do you any good.

Q & A

If DHEA levels begin to fall in the third decade, why don't experts recommend DHEA supplements for younger adults?

Since so little is known about the effects of too much DHEA on health, it's considered risky for this age group to supplement without at least having one's DHEA levels checked first. Some adults may still have normal or close to normal levels. In a 1996 editorial, noted French scientist Etienne-Emile Balieu admits it would be fascinating to study DHEA administration in younger adults (forty- to sixty-year-olds), particularly if they have low DHEA levels. But first "we need to know more about the effects of DHEA in older individuals," he cautions.

Every time researchers talk about promoting longevity it seems they recommend exercise. Can't I just take DHEA supplements and let my wife go walking on her own?

Sure you can. But DHEA, as your doctor will tell you, can't make your out-of-shape body into a muscle-bound physique on its own. (See Chapter 3 to learn how much DHEA might do for your muscle stores.) Why not try to gradually build up to a more active lifestyle. A recent report published in the American Heart Association's journal of *Circulation* finds that an aerobic exercise program for sedentary older adults improves cardiovascular function regardless of prior physical conditioning. In other words, the body never loses the ability to get back into shape. Edward Lakatta, M.D., chief of the Laboratory of Cardiovascular Science at the National Institute of Aging and primary investigator on the study, points out that the number of cases of heart disease and stroke rises steeply after the age of sixty-five, accounting for more than 40 percent of all deaths among people age sixty-five to seventy-four. But for active older adults, such as older athletes, the heart functions much like that of a younger man. If you want to live longer and healthier, there's no doubt that keeping active is an important part of the equation.

Didn't I read something about eating less food might be one way to lengthen lifespan?

Yes. Studies with animals continue to demonstrate that cutting calories by about half (mind you this is a lab rat not a human) can help lengthen lifespan as much as 50 percent. Several researchers are testing the concept on themselves but the bulk of antiaging experts are against calorie restriction as an antiaging tool. The reason is basic. Most people already require fewer calories as they age yet seem to have more difficulty getting enough of certain key nutrients necessary for health. Since many nutrients are being singled out as potential enhancers of immune function (see Chapter 7 for that list) and since nutrients such as calcium are needed to keep bones strong it seems unwise to cut back on calories.

DHEA

Does anyone know how big a role genetics might play in determining lifespan length?

Not exactly. But what researchers do know is that environmental factors—smoking, poor diet, inactivity—help to cut a lifespan short no matter what the genetic blueprint. There's no precise guarantee that employing healthful strategies like a low-fat diet, and perhaps replacement hormone therapy with DHEA or estrogen or testosterone (see Chapter 10) will make you live to eighty or ninety or beyond. But every indication is that taking care of your body will help improve the quality of your life no matter how long you live.

Chapter 3

Controlling Your Weight, Building More Muscle

In a world where the waist on Barbie dolls continues to grow smaller and male models sport lean physiques, it's not easy to fight the image that thinner is better. In fact, chances are you've tried some pretty wild schemes in the past to help shed extra pounds: "Lose While You Snooze," the "Last Chance" Diet, the "All-You-Can Eat Grapefruit Plan." The names aren't important. The point is that the effects of these kinds of weight-loss plans typically don't last. That's why it's so intriguing to think that a hormone such as DHEA might hold the key to helping keep weight off. Wouldn't it be nice to be able to take a pill that could help you shed body fat, boost muscle, and turn off an overactive appetite? Sound too good to be true? Some experts are convinced it's not. If you've ever battled the bulge you might want to hear what they have to say.

DHEA

A story about "the fatty rat"

Pinning down what causes some people to gain weight so easily while others have no problems controlling weight is no easy task. But over the last few decades studies with laboratory animals have helped to shed at least some light on the issue. Studies with one animal model—the "Zucker fatty rat" as scientists label it—have proven particularly valuable, mainly because the pattern of obesity in this animal shares many similarities with the development of obesity in people. For instance, these rats overeat, are underactive, have mild elevations in blood sugar, and can develop a resistance to insulin, the hormone that regulates blood sugar. It makes sense then that any study with Zucker rats could have implications for people. In fact, that's why researchers are so excited about the potential weight-control benefits of DHEA. Several different studies show that Zucker rats (as well as dogs and other animals) fed DHEA lose body fat and not surprisingly, body weight.

Does the hormone have some control over appetite; is it an appetite suppressant perhaps? Or might it somehow influence behaviors that lead to overeating? In a 1993 study with Zucker rats, scientists at Louisiana State University Medical Center in New Orleans were hoping to identify the antiobesity mechanism of action for DHEA. The researchers fed lean and obese Zucker rats either plain rat chow or DHEA supplemented chow for one month. Obese rats nibbling on the DHEA–containing chow ate far less than obese rats given the control chow. Strangely, lean rats dining on the DHEA chow ate more.

Results, say the researchers, suggest that there may be an appetite component to DHEA's antiobesity effect, at least in the obese animal model. It's too early to tell whether this finding might apply to people. The one big drawback to DHEA as a potential treatment for obesity is that it appears to take large quantities of the hormone to have an impact. That presents a real dilemma since presently, very little is known about continued use of even small doses of DHEA. (The longest study of DHEA supplementation to date lasted just six months.)

Potent combination: DHEA plus weight-loss drugs

Researchers at Louisiana State University, the same scientists who completed the studies mentioned above, think there may be a way around the high dose dilemma: combine DHEA with another weight-loss agent. They are testing a combination of DHEA and the prescription antidepressant fenfluramine on Zucker rats. Chances are you recognize the diet drug fenfluramine as half of the highly advertised "phen-fen" program being touted by weight-loss clinics around the country. As an appetite suppressant, fenfluramine has been on the market for several years and has helped many obese people to lose weight. Its downside, however, is that the drug can't be used for long periods of time since dieters acquire a tolerance to it. By combining fenfluramine with DHEA researchers think they may have hit upon a way to play up the positives of both substances and downplay the negatives. Preliminary findings suggest this pair of appetite suppressants works to promote weight loss in animals. In fact, by using DHEA researchers say it's possible to extend the time that fenfluramine can be used before a tolerance develops. More studies are needed to replicate these early findings and possibly pave the way for a future diet drug use for DHEA.

Can a pill build muscle and trim fat?

One of the more perplexing findings about the potential antiobesity effect for DHEA is that it appears to be sex specific. That is, men seem to benefit but women do not. In one of the earliest studies to notice that effect on men, researcher John E. Nestler and colleagues at the Medical College of Virginia in Richmond gave a small group of normal weight young men 1,600 milligrams of DHEA for four weeks and were able to produce an amazing 31 percent drop in body fat percentage. But later reports have shown mixed results. In fact, another more recent 1990 study from the Medical College of Virginia finds that when obese men are given that same DHEA dose, 1,600 milligrams, for four weeks there is no resulting change in body fatness. Baylor College of Medicine's Peter Casson reports just completing a month-long study with men in which participants were given a daily dose of 50 milligrams of DHEA. Drop in body fat percentage turned out to be a more modest 5 percent. Casson told professionals at a conference in September 1996 that the majority of studies seem to indicate that the antiobesity effect of DHEA, if it is there, appears to be mild.

That seems to agree with the findings from the very first clinical trials with DHEA supplements done by researchers at the University of California at San Diego. To repeat it again, these scientists noticed that men taking 100 milligrams of DHEA for six months lost roughly 5 percent of their body fat stores. Women did not show any significant changes even though they were taking the same dose of the hormone for the exact same amount of time. When muscle stores were measured, men and women both had a significant increase.

In a 1995 report published in *Fertility and Sterility* researchers testing the safety and efficacy of different doses of DHEA supplements (25 milligrams versus 50 milligrams) for postmenopausal women wondered if they might also be able to shed some more light on the obesity issue. Their results: Female volunteers, both those who were normal weight and those who were overweight, experienced no significant change in lean body mass (muscle tissue) or the percentage of body fat on either dose of DHEA. In fact, these scientists say that they have never noticed any kind of change in body fat levels in any of the many studies they've conducted with women.

Might DHEA dose be the issue? Remember that the studies by Dr. Yen and his colleagues (see Chapter 2), found no changes in body fat in either men or women after three months of supplementation with 50 milligrams of DHEA. It was when the supplement dose was increased to 100 milligrams and the time frame for supplementation was extended to six months that there was a small increase in lean body mass in both men and women.

Fat is still fattening

Even if future studies prove that DHEA, with or without the aid of other weight-loss medications, can help people shed excess pounds that won't mean that diet and exercise are no longer important. Slouching in front of the television and stuffing on high-fat snacks is a surefire recipe for weight gain regardless of what kind of diet pill you swallow. That's why diet clinics that promote the use of drugs also encourage people to change their eating habits. The reason is simple. Eating lots of fatty food can lead to a fatty body.

Utah scientists studying the diets of more than 200 healthy men aged twenty to seventy-one document this connection. After controlling for factors like age, aerobic fitness, and body weight, there was still a link between composition of the diet and the percentage of body fat. Men with the highest levels of body fat ate significantly more dietary fat and significantly less carbohydrate, complex carbohydrate, and fiber than men with the lowest percentages of body fat. Interestingly, the amount of calories each of the men ate wasn't all that different. Each of the men seemed to eat the number of calories needed to maintain weight. So it wasn't so much calories that seemed to be important but where those calories came from. Fat, in other words, was more fattening.

What is a "healthy" weight?

However, before you start worrying about DHEA and shedding pounds take a look at whether or not your current weight is unhealthy. It's definitely true that too many excess pounds may increase your risk for high blood pressure, heart disease, stroke, diabetes, and some types of cancer. But experts say a number of factors play into what constitutes a healthy weight: where your fat is stored, how much of your weight is muscle versus how much is body fat, and your overall health profile. This weight chart will give you a ballpark range of what weights are considered healthy based on age and height. *Note:* The higher weights in the ranges generally apply to men, who tend to have more muscle and bone; the lower weights more often apply to women, who have less muscle and bone.

Suggested Weights for Adults

Height (w/out shoes)	Weight in pounds (w/out clothes)	
	19 to 34 years	**35 years and over**
5'0"	97–128	108–138
5'1"	101–132	111–143
5'2"	104–137	115–148
5'3"	107–141	119–152
5'4"	111–146	122–157
5'5"	114–150	126–162
5'6"	118–155	130–167
5'7"	121–160	134–172
5'8"	125–164	138–178
5'9"	129–169	142–183
5'10"	132–174	146–188
5'11"	136–179	151–194
6'	140–184	155–199
6'1"	144–189	159–205
6'2"	148–195	164–210
6'3"	152–200	168–216
6'4"	156–205	173–222
6'5"	160–211	177–228
6'6"	164–216	182–234

Source: Nutrition and Your Health: Dietary Guidelines for Americans, third edition, 1990. U.S. Department of Agriculture, Department of Health and Human Services.

Figuring out how much to eat

If you've decided you need to lose weight and your doctor approves of experimenting with DHEA supplements you'll still want to watch what you eat. Limiting fatty indulgences such as ice cream, whole milk, fatty meats, margarine and butter is a good place to start. But don't make the mistake of thinking that if a food is fat-free you can eat unlimited quantities. In a study conducted at the University of Colorado, researchers put dieters on either one of two weight-loss plans. One group learned how to limit fat but were told to eat any quantity of starchy foods and other carbohydrates they desired. The other group learned to limit both fat and calories. Not surprisingly, women who counted calories lost twice as much weight (nineteen pounds versus nine) as the group that ate carbohydrates ad lib. Men who counted calories lost twenty-six pounds; their ad lib cohorts lost only eighteen. The health message here is simple: calories do count if you plan to lose weight.

Don't be an overzealous calorie slasher

It's uncertain if DHEA has any impact on metabolic rate. Still, whether you are taking DHEA or not, bear in mind that too stringent a diet can cause weight-loss efforts to backfire. If your body senses a food shortage it slows metabolism in an effort to burn less energy. To make sure you don't cut back too far on calories, consider calculating your energy needs with the formula used by professionals. It's not really that complex; the formula varies slightly for men and women. Bear in mind that the number you come up with is an estimate of the energy your body burns at rest. You'll need to add extra calories to this resting state to compensate for the energy you burn with exercise and activity.

MEN: Insert your age, height, and weight where appropriate. The example uses a forty-nine-year-old man who weighs 190 pounds and stands 5'11".
1. Multiply your height in inches by 12.7
   $(71 \times 12.7 = 902)$
   Multiply your weight in pounds by 6.3
   $(190 \times 6.3 = 1,197)$

2. Add these two numbers plus another sixty-six calories
   (902 + 1197 + 66 = 2165)
3. Multiply your age by 6.8. Subtract this number from the total in the step above. That will be your resting metabolic rate.
   (49 × 6.8 = 333 − 2,165 = 1,832 calories)

WOMEN: Insert your age, height, and weight where appropriate. The example uses a fifty-one-year-old woman who weighs 140 pounds and stands 5'4".
1. Multiply your height in inches by 4.7
   (64 × 4.7 = 301)
   Multiply your weight in pounds by 4.3
   (140 × 4.3 = 602)
2. Add these two numbers together plus another 655 calories.
   (301 + 602 + 655 = 1,558)
3. Multiply your age by 4.7. Subtract this number from the total in the step above. This will be your resting metabolic rate.
   (51 × 4.7 = 240 − 1,558 = 1,318)

Thinking and eating thin

If there is one lesson to be learned with weight loss, it's that patience and small changes pay off. Making little adjustments in how much you eat and how often you exercise may not sound as trendy as taking a DHEA tablet, but it does pay off. That's because the secret to weight loss isn't in any bottle or crazy diet scheme, it's with you. If you gradually change the health habits that made you put weight on in the first place, pounds will melt away permanently. Here are some tips on how to behave and become thin.

- *Eat breakfast, lunch, and dinner*

A study from the Mayo Clinic finds that breakfast skippers burn about 5 percent fewer calories over the course of a day than people who stick to the three or five meals a day routine.

- *Set a leisurely pace at meals*

It takes about twenty minutes for the stomach to send a message to the brain that it is full. Yet most of us devour meals in half that amount of time, leaving open a window of temptation or enough time to stuff in a lot more food than needed to satisfy hunger.

- *Unlearn your preference for fat*

In a 1993 study, researchers at Monell Chemical Senses Institute in Philadelphia found that volunteers given lower fat foods actually came to enjoy those selections and by study's end actually found higher fat foods less appetizing. It may be that the body acquires a "taste" for fat that can be unlearned in eight or twelve weeks.

- *Keep active*

Studies from Baylor College of Medicine in Houston, Texas, find that dieters who shun activity are more likely to regain weight in the two years following a diet. People who combine activity and sensible eating, on the other hand, lose weight steadily and keep it off during that same period of time.

Q&A

I've read that metabolic rate slows down after the age of thirty-five and that's why most people gain weight. Can DHEA do anything to prevent this change?

So far there is limited and contradictory evidence that DHEA has any impact on metabolic rate. But there may be another way you can circumvent this metabolic slowdown: boost your activity level and add strength or weight training to your fitness regimen. While past studies have confirmed that metabolic rate can slow down with age (probably about one-half of one percent each year) researchers are beginning to wonder if the change has more to do with inactivity than aging. When a group of Nevada researchers measured the metabolic rates of 500 men and women—ranging in age from twentysomething to sixtysomething—over the course of a five-year period they expected to see a slowdown as study participants aged. Interestingly, the numbers were across the board. The scientists could find no trend that was consistent with a metabolic slowdown based on age. In fact, current thinking is that one of the main reasons why metabolic rate goes down with age is because of diminishing levels of lean muscle mass. It all makes sense when you consider that it takes more calories to maintain muscle tissue than to maintain fat. Think of it this way: the more lean or muscle tissue you have, the more energy your body will burn no matter what your age.

Why do you suppose that DHEA has a mild antiobesity impact on men but not women?

Researchers would love to have that answer. It's obvious that there is an underlying trend of sex specific effects for DHEA, particularly with respect to obesity and diabetes. It's possible it can all be tied to biology (men have more muscle, higher testosterone levels). But right now scientists aren't even speculating what might precipitate the difference. When scientists uncover the mechanism by which DHEA works, it should help clear up the question of these sex-related differences. Incidentally, one recent research paper finds that dieting does have an impact on DHEA levels. But guess what? It occurs only in men. Apparently, men following a two-month low-calorie diet (1,000 to 1,400 calories) saw a 125 percent rise in serum DHEA levels. Women dieters experienced no change in DHEA although they showed similar amounts of weight loss.

Chapter 4

Protecting Your Heart: Men vs. Women

Although the death rate from heart disease continues to drop, probably due at least in part to advances in technology and the healthier lifestyle habits many Americans are adopting, this illness is still the nation's number-one killer. And as much as researchers know about cardiovascular disease, particularly atherosclerosis, the disease that clogs arteries and can shut off blood flow to the heart or brain, there are still quite a few more pieces to the puzzle. Preliminary evidence has some scientists wondering if DHEA may be one of those missing pieces, particularly for men.

Linking DHEA levels to heart disease

When reference is made to DHEA as a possible protective agent against heart disease, one of the major research projects typically offered in support comes from researchers at the University of California at San

Diego. Back in the early '70s researcher Elizabeth Barrett-Connor and colleagues recruited older residents of an upper-middle-class community near Los Angeles called Rancho Bernardo. The idea was to track a variety of different risk factors for cardiovascular disease. A whole battery of characteristics was measured including everything from blood pressure to blood sugar levels to DHEA. (Interestingly, DHEA levels weren't tabulated until the mideighties. It seems you can freeze blood samples taken from volunteers for up to fifteen years without affecting the concentration of DHEA.)

In the first Rancho Bernardo report published in 1986 these scientists noticed a correlation between lower DHEAS levels and increased heart disease risk among men. Since the study looked only at a small segment of the Rancho Bernardo Cohort, and since there seemed to be no apparent connection between DHEAS levels and heart disease risk in women, the scientists decided to test a larger segment of their study population to see if the trend would hold.

In that follow-up study there was a connection between lower levels of DHEAS and heart disease risk, specifically risk for heart attack. Again, there was no connection for women. What does this mean? Perhaps not as much as some proponents of DHEA replacement would like you to believe. There is a correlation between low levels of DHEA in the blood of men and an increased risk for heart disease. But it's taking a giant leap to assume that raising a low blood level of DHEA can help prevent heart disease. Yet, early results of studies with animals and population data about DHEA levels have created an intriguing new area of study for scientists.

Of clogged arteries and atherosclerosis

It's common knowledge that atherosclerosis, or the clogging of arteries due to fatty plaque buildup, sets the stage for coronary artery disease. But the underlying mechanism that leads to the development of atherosclerosis is less clear. Research does show that smoking and high blood pressure can damage blood vessel linings, offering an easier-to-adhere-to surface for fatty plaque. And fatty diets, particularly diets that include large amounts of saturated fat such as that found in butter, cheese, and well-marbled red meat, raise the level of cholesterol in the blood, offering plenty of raw materials for plaque formation. But the mechanism that triggers cholesterol to clump together with other particles and begin to plug up arteries remains elusive.

One theory that is rapidly gaining support among researchers ties at least some of the damage to chemical reactions in the body started by free radicals. Free radicals are produced daily by the body during numerous routine chemical reactions and are absorbed from environmental sources such as cigarette smoke and ozone. As we've seen earlier, speculation is that DHEA may play a role in neutralizing these free radicals. In order to understand the role scientists are proposing for DHEA, it's important to realize that free radicals are highly unstable substances that can wreak catastrophic damage if they are not successfully neutralized. Antioxidants in the body and in foods (vitamin E, vitamin C, beta-carotene) can help block some of the damage. They provide free radicals with an electron, thereby stabilizing the radical and preventing it from oxidizing or damaging tissue.

Speculation is that free radicals that interact with low-density lipoprotein (LDL) cholesterol particles floating in the blood alter these particles and make them "stickier" and more likely to adhere to blood vessel walls. If evidence to support this theory continues to build, that could mean blocking LDL oxidation could be a potentially powerful method for preventing or altering the course of atherosclerosis. Remember, blood vessels that gradually narrow due to plaque can eventually shut off blood flow entirely. When arteries supplying the heart become totally blocked, heart attack is the result. If arteries supplying the brain become clogged the result could be a potentially debilitating stroke.

In a University of Tennessee study conducted on rabbits (these animals readily develop atherosclerosis after heart transplant surgery due to rapid buildup of plaque in arteries) DHEA helped to slow the progression of atherosclerosis. In a study with young men, researchers found dramatic decreases in LDL after administering 1,600 milligrams of DHEA. But scientists say it's too early to tell if DHEA is capable of blocking atherosclerosis or if some other factors may be at play.

Insulin resistance, diabetes, and DHEA

Some scientists think the impact DHEA has on heart disease risk might be indirectly related to diseases such as diabetes and insulin resistance. You may have heard the term "insulin resistant" being bandied about a great deal recently (mostly in stories about weight loss). Understanding how the body controls blood sugar can help to clarify the role insulin resistance and diabetes may play in raising heart disease risk. To put it simply, the body controls blood sugar with the help of a hormone secreted by the pancreas called insulin. If the pancreas doesn't secrete enough insulin, sugar builds up in the blood and causes damage. This can result in what is called Type I or insulin-dependent diabetes and usually occurs early in life. Injections of synthetically manufactured insulin are given daily to help control blood sugar.

Much more common is Type II or what is referred to as adult-onset diabetes. In this illness the pancreas manufactures insulin but for some reason this insulin doesn't seem to work effectively. Speculation is that this change in insulin sensitivity can start at around the age of thirty and may be programmed by a master genetic blueprint. Whatever the cause, when both insulin and sugar are allowed to remain in abnormally high levels in the blood, this can damage body tissues. Are you at risk? A combination of factors increase your risk of developing adult-onset diabetes: family history, being overweight, or inactive, over forty years of age, female. Women are about twice as likely to develop diabetes as men. But for some unknown reason, diabetes is more damaging to the hearts of women than men. It may be due to the fact that at the same time that many women are developing diabetes (forty-five years old or older) estrogen levels are dropping, which also increases heart disease risk. It's also well know that high blood sugar levels can damage blood vessels in both men and women.

Add some vitamin E, prevent heart attacks

While studies looking at DHEA as a potential antioxidant are in the early phase of the game, experts have a lot of data on the influence of nutrient antioxidant vitamin E on heart disease risk. Two 1993 studies from Harvard Medical School find a protective effect for high levels of vitamin E in the diet (from supplements) against heart disease. Now a new report from the University of Minnesota finds that eating foods rich in vitamin E may be even better than taking supplements, particularly when it comes to warding off heart attacks. Scientists came to that conclusion after studying the diet and supplement habits of a group of 34,500 postmenopausal women for seven years. Researchers divided women into five categories ranging from the highest food intake of vitamin E (more than 10 IU a day) to the least. Results clearly indicated that women who ate the most vitamin E were 62 percent less likely to die of coronary heart disease than women who ate the least.

The glitch in this advice, however, is that boosting vitamin E intake while keeping your diet low in fat is no easy task. That's because some of the richest sources of this fat-soluble item are high-fat items such as vegetable oils, nuts, and seeds. Go ahead and try to boost your dietary vitamin E with the foods below; if that doesn't work, you might want to talk with your doctor about taking a supplement.

Some food sources of vitamin E

Food	Amount	Vitamin E (IU)
Sunflower seeds	1 oz.	21
Almonds	1 oz.	11
Sunflower oil	1 tbsp.	10
Safflower oil	1 tbsp.	8
Wheat germ	1 oz.	5
Avocado	1 medium	3
Mango	1 medium	3
Margarine, soft	1 tbsp.	3
Mayonnaise	1 tbsp.	3

For women only

If you're like most women, you probably glanced over this section on heart disease and didn't start reading until you saw this message for women only. Perhaps the thought of developing breast cancer scares you a lot more. But the reality is, heart disease claims five times as many female lives as cancer of the breast. And of the more than 520,000 Americans who die from heart disease each year, nearly half are women.

Unfortunately, the fact that women develop heart problems about seven or eight years later in life than men makes many women too complacent about heart disease risk. According to the American Heart Association, one in nine women aged forty-five to sixty-four has some form of cardiovascular disease. By the age of sixty-five, the number climbs to one in three.

Although the major risk factors that set the stage for heart disease—high blood pressure, high cholesterol levels, smoking, obesity, a sedentary lifestyle—are the same for men and women, two of these factors play a different role in women. Smoking more than doubles the risk of heart attack for both men and women. But women who smoke and take oral contraceptives have a thirty-nine-fold increase in risk of heart attack as women who do neither. Before menopause women naturally have higher levels of high-density lipoprotein cholesterol (HDL), the protective package that carries cholesterol away from your arteries. Total cholesterol measurements may not estimate heart disease risk as accurately for women because they don't account for HDL cholesterol.

However, women do still have one big advantage. The female hormone estrogen appears to confer strong protection against heart disease. Part of that protection appears to be related to blood lipids. Estrogen helps raise the levels of HDL or "good" cholesterol and lower the levels of the more harmful low-density lipoprotein. The problem is that after menopause, when estrogen levels begin to decline gradually, that protection is gradually lost. Some time after the age of sixty-five, a woman's risk of heart disease actually becomes almost equal to a man's. Estrogen replacement therapy may be the way to recapture that protection.

Putting together a program to lower heart disease risk

While it's way too soon to make recommendations about how much DHEA might help protect the heart, and since DHEA may not offer protection against heart disease to women, it might be wise to build your heart disease prevention plan with proven strategies such as the ones below. Once you've adopted this kind of protective plan of action you might want to talk to your doctor about trying DHEA supplements. If you're a woman, don't be surprised if supplements are discouraged since preliminary evidence suggests that DHEA could have an adverse effect on your lipid (fat) levels.

1. *Don't smoke*—Toxic substances in cigarette smoke damage the lining of blood vessels making it easier for fatty plaque to build up on artery walls. Smoking also makes the heart work harder; nicotine constricts blood vessels and raises blood pressure. The more cigarettes you smoke, the higher your risk of developing heart disease. *Note for women:* Smoking wipes out any natural biological protection you might obtain from the hormone estrogen.

2. *Limit fat and cholesterol*—You might want to consider having blood cholesterol levels checked. The most desirable level for blood cholesterol is a number less than 200 mg/dL. A level above 240 mg/dL is considered dangerously "high." If you fall into the range in between those numbers (200–239 mg/dL), experts consider that a "borderline" high risk. Experts recommend limiting your fat intake to 30 percent of your calories or less. The Department of Agriculture suggests you figure the number of grams of fat that provide 30 percent of calories in your daily diet as follows: multiply your total day's calories by 0.30 to get your calories from fat per day. For example: 2,200 calories $\times$ 0.30 = 660 calories from fat. Next, divide calories from fat per day by nine (each gram of fat has nine calories) to get grams of fat per day. For example: 660 calories from fat 9 = 73 grams of fat. Once you find your fat budget, read labels and keep tabs on that total. Or just try for these general strategies:

—Use fats (butter, margarine) and vegetable oils sparingly in cooking or at the table. Nonstick pots and pans can help. So can the use of herbs, fat-free liquids (broth, fruit juices, wine vinegar), and other seasonings.

—Choose two to three servings of lean fish, poultry, meats, or other protein-rich foods such as dried beans daily. Try to purchase meats labeled "lean" or "extra lean." Trim the fat from meat and pull the fatty skin off poultry to cut fat content even further.

—Choose skim or low-fat dairy products, preferably skim if you can adjust to the change. One cup of skim milk has almost no fat. One cup of 2 percent milk has five grams of fat, most of it saturated. One cup of whole milk has about eight grams of fat.

3. *Exercise daily*—if you're sedentary you are twice as likely to develop heart disease as someone who keeps active. The 1996 Surgeon General's report on physical activity and health recommends adults accumulate thirty minutes of moderate intensity activity on most, if not all, days. Consensus is that less intense activities such as gardening and ballroom dancing can be just as beneficial to the heart as jogging.

4. *Maintain a healthy weight*—The more overweight you are, the higher your heart disease risk. (See Chapter 3 to determine a "healthy" weight for your age and height.)

Q & A

My wife tells me that stress is probably the reason I have a high blood cholesterol level. Is this true, and if it is I'm wondering if taking DHEA will help compensate for any rise in cholesterol caused by a stressful occupation?

Yes, stress could be raising your blood cholesterol level. But a steady diet of rich fatty foods and an inactive desk job may also be part of the problem. You'll need to talk to your doctor to figure out how different environmental factors (diet, stress, activity level) are influencing your individual number. As to the second part of your question, preliminary studies do suggest that DHEA may help lower blood cholesterol levels for men. But no amount of any hormone, including DHEA, can compensate for continually poor lifestyle habits. In other words, be sure to make healthful changes in the way you eat and live along with any DHEA supplement your doctor prescribes. (See Chapter 12 for tips on how to diffuse your stress.)

Just for your information, stress can have a pretty strong impact on blood cholesterol. In a 1992 report in the *Archives of Internal Medicine* researchers documented a rapid rise in blood cholesterol—an average hike of 22 points—among young men completing a set of mentally taxing word and math problems. And these were no sissy math problems. Complex sets of one-, two-, and three-column addition and subtractions flashed onto a computer screen and as participants solved those problems the computer continued to rapidly up the ante, spitting out more difficult problems and speeding up the pace. During the testing a voice synthesizer barked out random incorrect and correct responses to distract the men. And to add more pressure, scientists dangled a financial reward ($20) for every correct response. One caveat: It's unknown if these kinds of stress-induced elevations in cholesterol or the stresses you are under at work can cause a permanent rise in cholesterol. Interestingly, scientists conducting this study noticed that cholesterol levels of volunteers remained high during the thirty-minute recovery period after testing.

I've been taking estrogen since the onset of menopause. Would adding DHEA to the program offer more protection against heart disease?

Probably not. So far the preliminary findings don't seem to suggest that DHEA supplements are heart protective in women. Estrogen appears to yield the bulk of protection. Indeed, a new study reported in the journal *Obstetrics and Gynecology* confirms that women who receive long-term estrogen replacement therapy (ERT) after menopause—five years or more of supplemental estrogen—are less likely to die from any cause than women who don't take supplemental estrogen. Researchers say that this lower mortality risk is primarily due to a reduction in cardiovascular disease among ERT users.

I'm a forty-nine-year-old man with what the doctor tells me is a dangerously high blood cholesterol level. Didn't I read somewhere that some foods can actually help lower cholesterol? How about DHEA?

Definitely for the first question. Probably for the second. Studies show that an ever-expanding list of foods might help lower cholesterol levels possibly due to the type of fiber they contain or other helpful plant chemicals. On that list: Barley, garlic, grapefruit, oats, prunes, rice bran, and wine. One more thing: the American Heart Association and the National Heart, Lung, and Blood Institute both say that up to four egg yolks per week are permissible on a low-fat diet. Interestingly, shellfish, another food that used to be considered taboo for people with elevated cholesterol levels, is probably not that bad for you to have on occasion, either. Recent studies indicate that these foods from the sea don't raise blood cholesterol levels as dramatically as once thought. Preliminary evidence suggests that DHEA may help lower your cholesterol level, too, although the findings don't seem consistent.

I'm taking low-dose birth control pills and am already concerned about how these might raise my risk for heart disease. Where does DHEA fit into the picture for me?

No one can say for sure. According to the National Heart, Lung, and Blood Institute the risks associated with using low-dose birth control pills are not really known. (Much of the information doctors have about birth control pills and heart disease risk comes from studies with higher dose hormones used in the past. Women who used these higher-dose contraceptives appear more likely to suffer a heart attack or stroke due to a higher risk of blood clots.) Neither are the long-term risks of taking DHEA. Since preliminary findings don't suggest a cardioprotective role for DHEA in women, you'll need to beef up your heart disease prevention program with other strategies: stop smoking, lose weight if you are overweight, keep your blood pressure under control, and eat a diet low in fat and rich in vitamins. (See the Appendix I for the DHEA Food Plan.)

Chapter 5

Revving Up Your Libido

Gathering up popular magazines, particularly specialty ones targeted at men and women, you'd think that most Americans are desperately in search of a better sex life. Titles run the gamut from: Seven days to a better sex life, Seven weeks to better sex (the long approach, perhaps for those who have a little more patience?), to Sex after thirty-five, Great married sex, etc. But a handful of researchers suggest that there may be a much simpler approach to boosting libido and improving sex life. The magic, desire-stimulating elixir they refer to is none other than the adrenal androgen DHEA.

At this point, most of the evidence is anecdotal. Yet, it's not that much of a stretch to think that an androgen such as DHEA could influence sex drive since studies clearly demonstrate that other hormones (testosterone and estrogen) play a role in turning on and turning off our sexual appetites. Remember, the body can convert DHEA into testosterone and estrogen compounds.

DHEA

The hormones of sexual desire

A large body of scientific evidence supports the theory that androgens—testosterone is a potent androgen hormone—help determine sexual desire in men. For instance, in castrated men or men who have underactive testes (hypogonadism), manipulating the amount of hormone replacement therapy can manipulate the extent of sexual desire. Research shows that when replacement testosterone is completely withdrawn these men experience a sharp drop in sexual interest and sexual activity. On the other hand, resuming testosterone therapy significantly increases the frequency of sexual thoughts and fantasies as well as rekindles sexual desire. Unfortunately, the underlying mechanism behind this hormone-desire link is still poorly understood.

Interestingly, research clearly indicates that the amount of testosterone circulating in the blood of a random sampling of normal healthy men can vary dramatically. Yet there is no indication that these individual variations correlate to individual differences in sexual desire. Many scientists suspect that there could be a threshold level, a point at which additional testosterone has no influence on sexual behavior. If a man has adequate enough testosterone to meet that threshold it's possible that external sources of testosterone (drug patches with the hormone, or supplements of DHEA that the body can convert into testosterone) may be unnecessary. To complicate the picture even further, preliminary studies suggest that estrogen, other hormones, and naturally occurring opiates might also help to regulate the intensity and frequency of sexual desire and arousal in men.

Unfortunately, the role that reproductive hormones and testosterone play in controlling sexual desire in healthy women is less clear. Research does show that varying levels of the female sex hormones progesterone and estradiol don't seem to have any effect on sexual desire in young women. However, in premenopausal women that have had their ovaries removed, diminished sexual desire can be restored with testosterone replacement (with or without estradiol) or estradiol alone. Yet, these findings are far from a complete picture. Part of the problem with studying hormones and sexual desire in women is that many researchers have used the menstrual cycle as a base for study. The cyclic nature of the twenty-eight days of a woman's menstrual cycle seems like it might be helpful in teasing out hormonal influences. After all, there is a distinctive ebb and flow of various hormones throughout the cycle. But some researchers suspect that this cyclic fluctuation in hormones may be influenced by changes in mood and energy levels, changes that vary dramatically from woman to woman. More importantly, many studies have failed to identify menstrual cycle-related fluctuations in sexual arousal.

Is testosterone then the more likely "libido-enhancing hormone" in women? Opinions are mixed. In the past, when scientists measured the level of circulating testosterone in women and then tried to relate those levels to sexual desire the results proved inconsistent. But there does seem to be a renewed interest in pursuing testosterone as the key to female libido. Susan Rako, M.D., a psychiatrist in private practice in Boston for the last twenty-five years, has written a book about her own research and experiences with testosterone replacement. In *The hormone of desire: The truth about sexuality, menopause, and testosterone,* she questions the resistance many doctors hold toward prescribing supplementary testosterone for a woman suffering symptoms of deficiency and reignites the debate about whether there is a link between the decreased testosterone levels that occur with menopause and diminished sex drive.

As researchers continue to tease out the influences of hormones on sexual appetite and behavior, keep in mind that libido is not solely influenced by chemical stimulation (aka hormones). In one recent 1995 report researchers conclude that it's difficult to try and "operationalize" sexual desire. Most likely sexual desire and sexual expression are a "complex interaction of hormonal effects, individual personality characteristics" and psychological and social factors based in some part on culture and experiences.

Sexuality and sexual desire: A complex picture

Most experts concur that every individual has his or her own distinct psychological and behavioral blueprint, a blueprint that factors into the libido equation quite heavily, perhaps just as heavily as hormonal influences. One report in the *Psychology of Women Quarterly* suggests that psychological response to menopause can affect sexual functioning. Although research seldom addresses this issue, this scientist suggests that the physiological symptoms that signal menopause also signal aging to many women; they may for the first time begin to realize their own mortality. If a woman has a family, this may be the time that children are growing up and leaving home, what is often referred to as the "empty nest syndrome." These major life transitions, and in some cases stressors, can influence sexual desire.

It may be no different for men. They may not experience the "empty nest syndrome" but are beginning to realize their own mortality, too. Then there's the negative impact from a whole host of psychological factors: excessive stress (not uncommon in the fast-paced '90s), anxiety, depression, unresolved problems or relationship conflicts with a sexual partner. Indeed, if the brain, a potent sexual organ, is tied up with other issues and conflicts this may cause a wane in sexual desire. As for sexual arousal, animal research shows that the brain is involved. The areas of the brain that have been implicated include a region in the hypothalamus and related hippocampal structures.

Finding ways to successfully navigate through these psychological issues may be just as important to sexual health as hormonal replacement therapy. Keep in mind that physical health also impacts on sexual health. A loss of libido can be symptomatic of any number of physical ailments including depression, drug abuse, and alcoholic dependencies. Medications such as high blood pressure pills and sleeping aids can also inhibit sexual desire.

Can restoring DHEA to youthful levels renew a lack of interest in sex?

Will DHEA fix all the physical and psychological issues that surround libido? Maybe. Maybe not. Sexual appetites can vary throughout life and the occasional lull in interest is not unusual. Yet if lack of sexual desire is a persistent problem it could be due to a hormonal imbalance. While there is a lot of anecdotal evidence to support that DHEA might enhance libido, so far the research results are inconclusive. Interestingly, in one of the first clinical trials done with DHEA supplements, participants reported improved feelings of "well-being" but no change in libido either while taking DHEA or an inactive placebo.

Most experts feel that the study conducted at the University of California at San Diego in which older men and women were given replacement levels of DHEA, fifty milligrams a day for three months, is a solid one. The fact that these researchers saw essentially no change in libido among trial participants during the DHEA and the placebo phase is evidence enough for many scientists that DHEA is not a libido-enhancing agent. More research may help settle this controversy.

Are medications turning off your libido?

Before you blame all of your libido problems on waning hormone levels (DHEA, testosterone and estrogen), take a look in your medicine cabinet. Any number of prescription drugs are known to have a negative effect on libido. Some over-the-counter medications, particularly ones that induce drowsiness, can also dampen desire. Researchers aren't certain how different medications exert their influence on sexual function. But there are enough controlled studies, clinical trials and case reports of individuals taking various drugs and experiencing sexual side effects to warrant cause for concern. Here's a brief overview of some of the sexual problems encountered with various categories of prescription drugs. It's followed by a comprehensive list of potential problem drugs, ones that can impact libido, arousal or orgasm. If you think any of these drugs might be influencing your libido, talk with your physician about alternative courses of treatment.

Prescription drugs

Tranquilizers—Although most people taking drugs such as Valium (diazepam) and Xanax (alprazolam) don't realize it, these mood-altering medications can depress sexual desire and cause problems with arousal and sexual response. Librium (chlordiazepoxide) is also on this list.

Antidepressants—A large number of antidepressants have been reported to cause several sexual side effects: diminished libido, impaired erections, orgasmic dysfunction. Some of the better known meds in this category are Prozac (fluoxetine) and Elavil (amitriptyline.) One caveat: it's possible that depression itself may be responsible for some of these sexual problems.

Blood pressure drugs—Loss of libido and problems with sexual function are a frequent side effect of drugs that lower blood pressure such as Inderal (propranolol), Aldomet (methyldopa), and Catapres (clonidine). In fact, sexual problems are so common that some practitioners assume drug therapy is responsible for any sexual problems that coincide with the new prescription of a blood pressure-lowering drug. That is, until it can be proven otherwise. Switching to a different drug may help. Or if you're overweight, you might try weight loss and exercise, two natural strategies that have been shown to help lower blood pressure.

Ulcer drugs—Medical advances show that many ulcers are due to bacterial infections. However, most physicians still use ulcer drugs such as Tagamet (cimetidine) along with antibacterial (antibiotic) therapy. Unfortunately, studies show that cimetidine can inhibit male androgen hormones. Men often experience decreased libido and have problems achieving an erection. The impact on female libido is less clear. Conversely, Zantac (ranitidine hydrochloride) does not appear to have this side effect.

DHEA

Q&A

Are DHEA levels the same in men and women, or is there a sex-related difference in DHEA production?

There are definitely some similarities as well as some differences between the sexes. Overall, experts find that there is a wide inter-individual range of DHEAS (up to threefold difference) among people of the same age regardless of sex. But women, as a group, tend to have lower DHEA levels than men. In addition, preliminary findings suggest that postmenopausal women who undergo estrogen-replacement therapy have even lower levels of DHEA than those who don't.

Estrogen replacement therapy has alleviated my hot flashes and I feel great. But my libido is about nil. I'd like to try testosterone or DHEA but I'm concerned about the masculinizing effects. Will I grow facial hair or a beard?

Talk to your doctor. It seems that only very high doses of androgens like testosterone and DHEA, doses taken for long periods of time, can produce masculinizing effects such as facial hair. In one study a participant did develop facial hair on fifty milligrams of DHEA taken over a three-month period. But the problem disappeared when supplementation was stopped. Many experts feel that the safer dose of DHEA for women is twenty-five milligrams per day. They also speculate that oral supplements may eventually be replaced with nonoral administration routes (transdermal patch, vaginal cream) as these routes might prevent some of the effects of androgenization. With oral administration, much of the DHEA you absorb is shuttled first to the liver where it is converted to potent androgens.

I'm on the "pill" and have always had the niggling suspicion that it influences my level of desire, not all the time but at certain points in my menstrual cycle. Can I take DHEA at these times to make up for the lack of desire?

Numerous studies do confirm that it is not uncommon for women taking oral contraceptives (OCs) to report a loss of sexual desire, so your suspicions may be founded. In fact, a 1993 study from researchers in Canada and Scotland found that women given triphasic OCs consistently reported decreased sexual interest during the menstrual and postmenstrual phases of their cycle; women taking a placebo did not report this loss. While DHEA can be converted by the body into testosterone, and some scientists think adequate testosterone levels may be part of the key to female libido, it's best to talk to your doctor before adding DHEA to your hormone regimen. Too little is known about how this hormone acts in women. Some scientists speculate that DHEA could have a negative impact on blood lipids (fat) and thereby increase risk for heart disease. Talk to your doctor about the potential risks and benefits.

I'm fifty-five and don't think about sex or have sex as often as I did at twenty-five. But I don't consider myself abnormal or unsatisfied and my wife is not complaining, either. If I take DHEA to prevent heart disease or for other health reasons is this going to shoot my libido into overdrive?

Probably not, at least not if the doses you take are small. But that's something you should discuss with your doctor. He or she can start you out on a small dose of DHEA and follow your blood levels until you achieve a point that is beneficial to you. By the way, you're not alone when you say you don't think about sex as often as you used to. Since the early eighties

DHEA

Money magazine has been polling Americans to figure out what occupies their minds more, money or sex? Surprisingly, in one of the latest surveys (1995) the interest in both areas seemed to be dropping considerably. About 44 percent of people report that they think more about money than sex; that's down from 47 percent in the 1994 survey. A paltry 15 percent say they think more about sex than money, a drop of two percent from the year before. And get this—a growing number of folks (16 percent) swear they don't contemplate either!

Chapter 6

Lightening Your Mood, Increasing Your Energy

It's common to blame some of the changes in mental capacity that occur as people grow older on aging. But over the last decade or so researchers have come to realize that some of these cognitive changes are not necessarily an inevitable physiological event. To some extent, the mental decline most of us assume is a natural part of aging is simply a case of lack of use, a "you didn't use it, so now you're going to lose it" situation. Speculation is that it may even be possible to reverse some of the cognitive deterioration that occurs with age, a deterioration that in many cases has more to do with environmental factors (activity, lifestyle) than with any kind of irreversible biological change. The last ten or twenty years have also brought a better understanding of the chemical milieu of the brain. Scientists now realize that certain brain chemicals, some of which are called neurotransmitters (serotonin, endorphins) can influence mood, thinking, and perhaps memory. Many suspect that the adrenal hormone DHEA may be on that list of influential brain chemicals.

Can DHEA be manufactured by the brain?

At an international meeting in Rome in 1989, French researcher Etienne-Emile Balieu presented evidence that DHEA can actually be synthesized in the animal brain. Cells in the brain called oligodendrocytes, the same cells that also manufacture the protective myelin coating that surrounds nerve fibers, are the site of production. And just as with adrenal production of steroids, these brain cells manufacture DHEA from cholesterol.

Balieu and colleagues suspect that these findings in the animal brain are "probably applicable to the human species." What's really exciting, however, is the very fact that this is happening, that the brain actually has the capacity to produce steroid hormones such as DHEA. That could be a good indication of their importance in brain function, something that human clinical trials may one day prove.

Improving well-being: more energy, a better mood

The study that many point to as an endorsement of DHEA's mood- and energy-boosting skills is one of the first human clinical trials with DHEA supplements. Conducted at the University of California at San Diego, researchers Morales, Nolan, Nelson, and Yen gave fifty milligrams of DHEA to a group of older men and women for a period of three months. While receiving supplements participants reported increased feelings of well-being. That included a better mood, a better ability to handle stress, and more energy. These older men and women also commented that the quality of their sleep was noticeably improved, something that a recent German study seems to duplicate.

In the German study, researchers noticed that volunteers given megadoses of DHEA (ten times the replacement dose of fifty milligrams) before retiring for the evening had increased levels of REM or rapid eye movement sleep. REM sleep is the cycle of sleep in which dreams occur. Speculation is that better quality REM sleep may lead to improvements in memory.

Is it memory loss or Alzheimer's?

That DHEA might improve memory is an enticing prospect, particularly in light of the fact that most of us experience some amount of memory loss as we get older and fear that the onset of forgetfulness might be a sign of a deteriorating mind. Yet, experts say not to sweat this small stuff. Normally, the brain works somewhat along the lines of a supersophisticated computer. You might liken it to a 140 billion megabyte system since there are roughly 140 billion nerve cells (called neurons) that relay information throughout the brain and the body (via the central nervous system) by either chemical or electrical signals. Occasionally, these circuits can get overloaded at any age.

When does this memory loss signal something more serious, such as dementia or Alzheimer's disease? That's difficult to say. Scientists know there is a genetic component to Alzheimer's and that the hormone estrogen may also play a role in dementia. But other than that, researchers are pretty much in the dark about what other factors, including environmental influences, might cause or help prevent this devastating illness.

Figuring out the genetic component

In a recent report, a group of Boston researchers wondered if published accounts that Alzheimer's is destructive to the minds of, and eventually kills, roughly 40 percent of people aged eighty-five might not be greatly overestimated. They point out that many of the studies used as a basis to form those predictions didn't include subjects older than ninety-three. Why include older subjects? you might wonder. Wouldn't they be more likely to be cognitively impaired? Not necessarily. When studying a small group of centenarian residents of a Boston aging center, these researchers found that only four out of a group of twelve seemed to have signs of Alzheimer's disease. They propose that specific genes may determine not only the biologic rate at which organs like the brain age, but also how well the body copes with Alzheimer's or any disease.

Two basic concepts could be at work. Genes might determine what researchers refer to as your "adaptive capacity," your ability to recuperate from any illness or fend off disease. Genetic blueprints could also influence your "functional reserves." That is, they may program the rate at which a certain body organ ages over the course of the lifetime and how much of that organ is needed to function properly. So what does this all mean in terms of Alzheimer's? Harvard geriatrician Thomas T. Perls, in an article for *Scientific American,* explains it this way. "The importance of these two characteristics to the survival of many oldest old can be seen in the varying effects that the buildup of neurofibrillary tangles has on cognition. Neurofibrillary tangles describe the web of dead brain cells that occur naturally with aging but appear in abundance in patients with Alzheimer's disease. The number of tangles that can accumulate before signs of Alzheimer's disease emerge varies among patients."

Perls presents an 103-year-old gentleman who displayed few outward signs of Alzheimer's disease as an example of these differences. After the man's death an autopsy revealed that he had a number of neurofibrillary tangles, tangles that in a younger brain would have suggested a probable diagnosis of dementia. Speculation is that this older gentleman had enough reserve brain function that he could compensate for the devastating damage already done to major parts of his brain. It's entirely possible that some people could have a delayed buildup of tangles and when these tangles do occur, a high tolerance for them. Those two positive forces could allow a person to remain mentally sound well into old age, showing no obvious signs of Alzheimer's.

Can DHEA help treat depression?

Scattered reports indicate that DHEA levels are lower not only in people with Alzheimer's disease but also in those with major depression. In addition, research confirms that patients with psychiatric disorders, particularly those with depression, often have all kinds of alterations in neuroendocrine function, that is, changes in the HPA (hypothalamic-pituitary-adrenocortical) communication network. In light of these findings and the fact that one clinical trial—the Morales and Yen supplement trial mentioned above—already suggests that DHEA supplements might elevate well-being, a group of California scientists decided to test and see if these benefits might apply to someone suffering with major depression. Very little research has looked into DHEA as a treatment for this disorder so researchers started out trying to answer a basic question. Might DHEA supplements improve mood and cognition in middle-aged to elderly (fifty-one to seventy-two years) people with major depression? The first pilot study was tiny, only three men and three women participated.

During the study, each of the volunteers was given thirty to ninety milligrams of oral DHEA for a period of four weeks. (Levels of supplements prescribed varied as researchers were attempting to offer enough replacement DHEA to return blood DHEA levels to a youthful state.) Each week, researchers conducted a battery of psychological tests including several tests that measure depression and tests of verbal memory. After the first study was completed, scientists singled out one participant, a sixty-seven-year-old woman, and followed her for another six months. She was given sixty milligrams of supplemental DHEA for four months; dosage was then increased to ninety milligrams for another two months. The results of the 1995 study, published in the *Annals of the New York Academy of Sciences,* indicate that as DHEA levels increased due to supplement use, symptoms of depression improved. The case was not as clear for memory. In the first part of the study, each of the six participants saw improvement in autonomic (involuntary) memory functions but other aspects of memory performance remained unchanged. However, in the second part of the study with the longer administration time (six months) other measures of memory improved. Interestingly, all noticed benefits ceased after DHEA was gradually withdrawn.

The downside of this first study is that supplements were given "open-label," which means participants knew what they were getting. Still, researchers were intrigued enough by these preliminary results to go one step further. Currently, larger-scale, double-blind clinical trials are underway in California and also in France to confirm and explore the impact of DHEA on mood. It will probably take a few years before researchers can amass enough data to draw any firm conclusions.

Are DHEA levels tied in some way to overall mental ability?

Realizing that DHEA levels can be lower in some people with Alzheimer's-type dementia and sometimes lower in people who are depressed it makes sense that scientists might want to step back and look for the larger comparison: Could DHEA be linked to cognitive ability? If DHEA levels decline with age and if mental abilities often diminish with age, are the two necessarily linked? Well, Drs. Barrett-Connor and Edelstein of the University of California at San Diego decided to test this concept.

Barrett-Connor and Edelstein measured the DHEAS levels of 270 men and 167 women. Then they tested cognitive function with five standard screening tests that can give scientists a comprehensive overview of your mental abilities. What happened when researchers compared blood levels of DHEAS to test scores? Not much. There appeared to be no connection between DHEAS levels and performance except for a single-memory association test in women. The test measured long-term (one minute) recall ability and verbal fluency (number of animals named within one minute). Scientists concluded that this one barely significant connection, in women but not in men, was likely due to chance rather than a true effect. However, they aren't ready to rule out a connection between DHEA and mental functioning until more studies are completed.

Keeping mindful of other hormones

In the meantime, keep in mind that DHEA doesn't appear to be the only hormone with a potentially potent impact on mood and perhaps mental capacity. Several studies now confirm that estrogen and testosterone may enhance cognitive skills. The bulk of the research to date has centered on estrogen, not because it has a more powerful impact on the mind, but simply because estrogen replacement therapy is more common and the more frequently studied of the two therapies. For example, in 1994, researchers at the State University of New York in Buffalo noticed significant improvement in mental ability in a group of older women receiving estrogen supplements. Scientists completed a battery of tests—memory, eye/hand coordination, problem solving with new information—on a group of thirty-six postmenopausal women both before and after they started taking supplemental estrogen. Mental scores improved across the board. The gains were subtle, say the researchers, but statistically significant.

In another unrelated report, women who had used estrogen for twenty years or more did much better on tests of verbal fluency than counterparts who had never taken estrogen. The most recent report on cognition and ERT, published in the *Journal of the American Geriatrics Society,* adds evidence that another female sex hormone, progesterone, may be involved in helping to maintain mental function. In this study, researchers had a group of 214 women residents of Leisure World Laguna Hills, a retirement community near Los Angeles, California, "draw the numbers on a clock face making no erasures." It sounds rather simplistic. Yet, scientists say that clock-drawing tasks are a good screening tool for cognitive impairment. An inability to position numbers on a clock correctly could be an early signal of cognitive decline.

When scientists compared clock drawings (168 women drew normal clocks; forty-six drew abnormal or blank clocks) they found no significant connection between estrogen use and clock-drawing performance. However, they did notice lower cholesterol levels and lower serum progesterone levels among women who drew normal clocks.

Testosterone for the mind

In unpublished preliminary reports from Johns Hopkins Medical School in Baltimore researchers are finding a connection between the primary male sex hormone, testosterone and learning ability. In a study presented at the International Congress of Endocrinology (June 1996) these scientists discussed the learning differences that occurred when a group of ten men with low testosterone levels were placed "on" and "off" testosterone replacement therapy. The study was broken into two phases; men were given an array of learning tests to determine such things as word, memory, and coordination skills during both segments.

Interestingly, while taking testosterone supplements men scored better on tests measuring visual and spatial skills. They had a better "mental grasp of objects" as evidenced by an improved ability to identify pictures and remember shapes as well as to fit geometric objects into the correct spaces. Conversely, when in the "off" testosterone phase, these same men were much more fluent with words and verbal memory. That is, they were better at constructing sentences and defining words. Researchers say there is still much to learn about how hormones influence cognitive skills and point out that studies on testosterone are probably a good fifteen years behind those of estrogen.

More ways to improve mental performance

Even if research continues to support the belief that DHEA and other hormones preserve or help improve some aspects of cognitive ability, they shouldn't be your only weapon against a deteriorating mind. Research clearly demonstrates that some of the cognitive changes that people experience as they grow older have more to do with disuse than dysfunction. Studies with one group of older adults clearly indicate that challenging the mind can actually help reverse mental decline. In fact, there are any number of steps you can take to keep your mind sharp and your memory youthful.

- **Consider some retraining**—For more than three decades scientists at Pennsylvania State University have been tracking the mental function of more than 5,000 men and women as part of the Seattle Longitudinal Study. Interestingly, they can find no uniform pattern of age-related changes in intellectual ability. More important, they have been able to help older adults regain lost cognitive skills with a simple training program. For example, in one study, a group of seventysomething women relearned the skills needed to read a road map, a kind of brushup course in spatial orientation. They also practiced their inductive reasoning by reviewing how to read and interpret timetables such as a train schedule. These might seem like simple challenges, but two out of three "students" showed noticeable improvement in skills after completing the training. A full 40 percent of the group improved so much that their scores reflected a level of functioning equivalent to the levels that they tested at fourteen years earlier when they first entered the study. Perhaps even more surprising, seven years later, these same women were still scoring better on cognitive skills tests, leading researchers to conclude that they were still experiencing the positive effects of training. Women who weren't enrolled in the training classes continued to experience a decline in mental function.

- **Employ mind-sharpening strategies**—Research also suggests that giving your brain a challenge on a regular basis stimulates the mind to function more efficiently. Some scientists liken the brain to a "mental muscle" that must be exercised in order to stay toned. How do you keep the mind challenged? Try crossword puzzles and anagrams. Play mind-stimulating games such as Scrabble or Boggle. Do anything that will keep the mind active and require you to think. One caveat: Don't count television shows as mind-stimulating. Experts say television watching is a passive one-dimensional activity. Educational and newsmagazine-style shows are the exception to the rule.

Hard as it might seem to believe, scientists have shown that the cerebral cortex, the area of the brain involved in conscious throught, can increase in size when laboratory animals are spurred by mental challenges. In what is now considered a classic study in the field of brain and cognitive research, researcher Marilyn C. Diamond of the University of California at Berkeley placed laboratory rats of varying ages in environments littered with stimulating learning toys and challenging exercise equipment. It didn't matter if the rats were young, middle-aged, or the age equivalent of Methuselah. The cerebral cortex grew larger in all of the mentally challenged animals. Diamond says the same scenario can apply to people.

- **Exercise regularly**—Salt Lake City researchers find that active men score far better on tests that measure key mental function—the ability to think quickly and shift attention rapidly from one thought to another, than men who are sedentary. The ages of participants ranged from twenty-one to sixty-two; higher activity levels equated with higher test scores. Will exercise work as well for a recovering couch potato? Definitely. These same researchers took inactive men (aged fifty-five to seventy) and put them in a walking program for four months. Every single one of the participants showed dramatic improvements on tests that measured cognitive and mental function.

- **Keep blood pressure under control**—Blood pressure is often referred to as the "silent" disease since people rarely suffer any overt symptoms. However, it's well known that blood pressure increases with age and now research shows that this increase could interfere with mental function. Scientists from New England find that people with high blood pressure score more poorly on tests of memory than their counterparts with normal blood pressure readings. Fortunately, the damage appears mild and is reversible.

Q & A

Is a study that suggests DHEA supplements improve well-being considered proof of a mood-enhancing benefit?

Not really. Yet, the fact that this study was a carefully done double-blind placebo-controlled endeavor makes these preliminary findings provocative. Other researchers have already taken the next step and are trying to see if the results can be duplicated. At a recent medical conference, Baylor College of Medicine's Peter Casson, M.D., mentioned that he and his colleagues are conducting a yearlong study of supplement use in men and women. Each participant is undergoing a comprehensive battery of psychological and cognitive testing. It's too soon to report the findings but Casson hints that early results in men show no mood-enhancing benefits. Results in women haven't yet been tabulated.

Even if DHEA may be helpful, isn't it true that genetics plays a big role in whether or not someone develops Alzheimer's?

Yes, and scientists are working diligently to untangle the genetic picture. A recent finding suggests that one variation of the gene coding for a protein called apolipoprotein E (*apo-E*) is tied to an increased risk of developing Alzheimer's. Apparently, there are three forms of *apo-E* genes that a person can inherit: *E2, E3,* and *E4.* If you happen to inherit two *E4* genes, one from each parent, studies show that your risk of developing the disease is eight times higher than that of the general population. Incidentally, these genes appear to program the age at which symptoms begin developing; sixty-eight is the average age of disease onset in people with two *E4* genes. If you carry two *E3* genes you also seem more likely to fall victim to Alzheimer's but at the much later age of seventy-five. Preliminary evidence suggests that *E2* genes may be protective against Alzheimer's, but their precise function is less certain. Researchers caution that more work needs to be done to determine the value of *apo-E* genes in predicting Alzheimer's.

Didn't I read somewhere that an aspirin a day might help ward off Alzheimer's? Would this be a better strategy than taking DHEA?

Not exactly. Studies show that people who routinely use nonsteroidal antiinflammatory (NSAIDs) drugs to treat their arthritis are less likely to have Alzheimer's than people who don't take these medications. And one small report finds that people with Alzheimer's who are given an NSAID called indomethacin do better on memory tests than folks who are given a placebo. But so far no one has directly tested aspirin, which is a far weaker antiinflammatory agent than any of the NSAIDs, as a potential preventive strategy for Alzheimer's. It's not impossible that some benefit might be found for aspirin in treatment of this disease but speculation is that the quantities of aspirin needed might be too excessive. Not much more is known about DHEA.

Chapter 7

Supercharging Your Immune System: Your Best Weapon Against Illness

Over the last few decades researchers have made dramatic advances in understanding the immune system. They've gained practical insights into how the body fends off illness and recuperates from injury, insights that support the concept that the weakening of immune response as one grows older is not inevitable. In fact, many scientists feel it's possible for people to take steps that will help strengthen internal defenses. In other words, you may be your own best weapon in the battle against everything from the common cold to more serious illnesses such as cancer.

Research is still in the early stages. But from vitamins to exercise to DHEA, the latest reports hint that with the right kind of lifestyle program you can harness internal disease fighters and take at least some control over the condition of your health. Even more astonishing, new research suggests that it may be possible to rejuvenate an aging immune system, one that is battle weary. To understand the significance of the latest immune/DHEA research, a quick review of how the body defends itself should prove helpful.

Immune defenses 101

Although the immune system is a complex network of cells and chemicals that is scattered throughout the body, it's mission is a simple one: to protect and to heal. Each link in the network is programmed to guard the body against poisons, microorganisms, foreign particles, diseased cells, or any kind of outside invader. To accomplish that job requires an elaborate system of checks, balances, and chemical reactions, some of which scientists understand but much of which still remains a puzzle.

Based on the current state of knowledge the body mounts several different types of defenses. Initially, the skin and mucous membranes form the first line of protection as they work to shut out invading microorganisms at the site of entry. Providing the central internal, or second line of defense, are white blood cells called lymphocytes. Primary warriors in the battle against infection, lymphocytes are produced in the bone marrow along with red blood cells and come in two varieties: T-cells and B-cells. What follows is a partial listing of some of the immune system components that may be influenced by DHEA.

T-cells—immune cells that primarily skirmish with viruses and parasites and may be involved in combatting cancer. T-cells fall into two main categories, either they are suppressor (killer) or helper cells. Each of these categories also have several components. For instance, a subset of helper T-cells are the CD4+ group. You'll read more about these helper T-cells in the chapter on autoimmune disorders (see Chapter 8).

T-cells derive their entire name from the fact that they have trained in the thymus, a gland situated in the upper chest that converts lymphocytes into T-cells. Helper T-cells typically detect the enemy and then signal killer T-cells to destroy it or call on B-cells to produce antibodies against the foreign substance.

B-cells—destroy bacterial invaders mainly by producing antibodies that fight these disease-causing agents. These antibodies lock on to the offending bacteria (antigen) before it can do damage to cells.

NK cells—natural killer cells; these immune warriors are a sort of surveillance patrol. Immunologists suspect that NK cells roam the body searching out wayward (malignant) cancer cells and viral invaders. Interestingly, NK cell production does not decline with age. DHEA has been shown to increase the number of NK cells in postmenopausal women.

Cytokines—hormonelike proteins secreted by various immune cells including the T-cells and macrophages. Two of the better-known classes of cytokines include the interleukins and the interferons, substances being tested as potential drug therapy in a number of different diseases.

Interleukins—chemicals secreted by the immune system's T-cells. Their role is to help recognize and mount an attack against foreign invaders. There are a handful of different interleukins with different functions, hence the names interluekin-1 (IL-1), interleukin-2 (IL-2), interleukin-6 (IL-6), etc. IL-2 appears to signal other interleukins and cytokines such as the interferons to activate the production of T-cells and B-cells in an effort to marshal the forces necessary to mount an attack against an invader.

Macrophages—jumbo-size immune cells that kill and engulf foreign invaders per instructions of T-cells and B-cells. Some macrophages maintain a home base (the spleen, the bone marrow, the liver, the lungs) and gulp down whatever foreign organisms happen to float by. Others travel the body hunting down foreign prey.

From "over the hill" to youthful vigor

Researchers refer to it as "immunosenescence" but in plainer language it refers simply to the aging of the immune system, and the good news is that it may not be inevitable. Studies with animals find that some of the changes that occur in immune function may be reversed by administering DHEA. The most exciting work comes from researcher Raymond A. Daynes and his colleagues at the University of Utah School of Medicine and the Geriatric Research, Education and Clinical Center in Salt Lake City, Utah. Daynes and fellow scientists provided aged animals with DHEA supplements (two to four milligrams of DHEAS per kilogram of body weigh; 2.2 kilograms is equivalent to one pound) and within days noticed that measures of immunocompetence were restored to youthful levels. More importantly, these older animals regained the strength to react to immune challenges in a competent fashion. Other researchers have confirmed these findings.

The animal evidence is so tantalizing that many immunologists are now interested in studying the effects of DHEA. Many scientists feel there is good reason to expect that human immune systems may react just as positively to DHEA supplementation. Some of the first human studies will look at how DHEA supplementation works in combination with vaccines.

The double whammy: DHEA plus vaccines

Another way DHEA may help the immune system is by boosting the effectiveness of vaccines. Normally, the immune system is capable of mounting a defense against any kind of foreign invader—bacteria, virus, toxin—by releasing antibodies that recognize the harmful substance and destroy it. However, the acquisition of immunity in response to an infection takes a few days or sometimes a few weeks to develop. During that time period the person is still going to be sick with the infection and so there is the potential that the disease will take its toll before immunity can be acquired. Luckily, scientists have been able to develop vaccines that help the body protect against some types of infections.

The way a vaccine works is fairly straightforward. Scientists develop harmless clones or in the case of influenza, a mixture of strains of the dead virus. Immune cells exposed to the vaccine learn to recognize the character of the invader so that when and if they do encounter it they can immediately destroy it. In other words, there is no need to wait a few days or weeks to "acquire" immunity, immune cells have mapped out their defense strategy in advance. (The incidence of quite a few diseases such as measles and diphtheria has dropped dramatically since the introduction of vaccines and routine immunization programs. One disease, smallpox, has been virtually eradicated.)

Studies clearly show that immune responsiveness can grow sluggish as people grow older. And studies show that response to vaccines is also weaker. Most older adults still have a reliable "immunological memory" for foreign substances encountered in the past. But when faced with a new foreign substance (antigen) the immune system often fails to elicit a response or only responds weakly. But researchers at the University of Utah may have found a way to turn the sluggish behavior around. When they administered DHEA along with vaccines in aged mice, response to new antigens increased dramatically. In studies with older volunteers over the age of sixty-five, this same kind of regimen—DHEA plus vaccine—showed mixed results.

Oral DHEA supplements given before administration of an influenza vaccine showed increased immune responsiveness as compared with placebo and vaccine. In a second study with tetanus vaccine, DHEA didn't appear to improve immune responsiveness although it didn't appear to interfere, either.

DHEA as healer: Amazing recovery from burns

If DHEA can boost immune function and possibly prevent illness, it seems plausible that this hormone might also play a key role in helping the immune system as it performs another key task: healing an injury. At least that was the hypothesis put forward by a group of scientists from the University of Utah School of Medicine. A few years ago these scientists began testing immune responsiveness in animals with burn damage. (Researchers first anesthetized the animals and then scalded a small amount of tissue on the animal's back to provide a similar amount of burn damaged tissue.)

Normally, immune function is weakened after a burn injury. Studies confirm that both animals and people are more susceptible to any kind of infection after a serious burn. But in this study, animals given an injection of DHEA in the days after burn damage actually showed an improvement in immune function. Specifically, treatment with DHEA helped maintain immune competence by preventing changes in the capacity of T-cells to produce other immune-protective agents, losses in cellular immune responses, and alterations in the ability of animals to resist an infectious challenge.

Yet, what proved even more astounding was something researchers hadn't even planned to study: the noticeable variation in the extent of tissue damage between the treated and untreated groups. When visually examining the burn sites of animals given DHEA, the extent and severity of tissue damage postrecovery was much less in the DHEA group. That led the scientists to wonder if DHEA might be exerting its protective effect by doing more than preventing changes in immune function. Was another mechanism at work helping limit tissue damage from the burn? Apparently, yes. In a second study done in 1995, researchers administered DHEA and some related steroid hormones under the skin of burned animals and then closely monitored tissue changes at the injury site, particularly the tiny capillaries and blood vessels that supply blood (oxygen, nutrients) to the damaged tissue.

After injections of DHEA most of the tiny dermal capillaries and venules within burn-exposed tissue maintained their normal "architecture." That is they stayed open. Researchers propose that DHEA, either directly or indirectly, helps maintain a normal architecture of dermal capillaries in the burn-damaged tissue and so may be useful in preventing the progressive tissue destruction caused by progressive ischemia. (Ischemia is the descriptive name given to the process whereby tiny clots develop in the vessels supplying blood to the injury site; as blood supply is totally cut off circulation comes to a standstill and eventually tissue in the damaged area dies.) Specifically, in one study lab mice received either an injection of 100 micrograms of DHEA (dissolved in a fluid medium of propylene glycol) or plain propylene glycol. In the DHEA group at sixteen days postinjury, the burned ear tissue had healed completely. Mice not receiving DHEA lost more than 70 percent of ear tissue.

Researchers continued with the experiments and tested the concept of the timing and amount of the dose of DHEA needed to be effective. Results showed that if DHEA was given up to four hours after injury, the beneficial effects on preserving tissue remained. At six hours it was no longer as effective. To test the dose, mice were given four different amounts—100, 50, 25, or 10 micrograms—on a daily basis until the experiment was concluded. The 10 microgram per day dose was not able to offer the same level of protection as the higher doses.

The mechanism has yet to be demonstrated, and researchers acknowledge that intervention with DHEA cannot prevent direct damage caused by the burn. But their findings, they say, strongly support the conclusion that DHEA "can partially, if not totally, prevent the progressive aspect of burn-induced ischemia of the skin." Moreover, this preventive capacity is dose-dependent but can be achieved even if therapy is delayed for several hours after the injury occurs. Speculation is that DHEA may be acting as an anticoagulant or a powerful antioxidant.

HIV: The DHEA-cortisol connection

As much as researchers still have to learn about human immunodeficiency virus (HIV), there is one factor they are certain about. Virus particles are capable of directly invading and damaging immune cells and are a major cause of weakened immune competence in people with the infection. However, experts aren't certain these virus particles account for all the immune damage. For instance, it's well known that certain steroid hormones, particularly the glucocorticoid hormone cortisol, can dampen immune response. Hope is that by piecing together the puzzle of how hormones such as cortisol and DHEA impact on immune function, researchers may be able to latch on to a treatment for this devastating virus, or at the very least be able to forestall some of damage it wreaks on immunity.

Here's a brief rundown of what scientists have uncovered so far that might help to establish a DHEA-cortisol-HIV link:

• Although adrenal glands are not severely damaged, several researchers have noticed mild impairment in adrenal function among people with HIV. That could influence production of adrenal hormones either directly, or perhaps indirectly due to deficiencies in some of the enzymes that produce these steroids.

• Scientists have documented elevated levels of glucocorticoids (cortisol) in people infected with HIV.

• One small study of thirteen HIV-infected patients reported lower levels of both DHEA and DHEA sulfate; levels appeared related to the severity of the disease.

• DHEA levels are low in people with acquired immunodeficiency syndrome (AIDS).

It's not certain what all these preliminary findings mean. One promising theory holds that DHEA and cortisol, both hormones made by the adrenal glands, could

DHEA

be an immune-controlling pair working in tandem to keep immune response balanced. If levels of the two hormones get too far "out-of-whack," this could precipitate major problems in mounting an immune defense. Keep in mind that studies seem to suggest that cortisol can suppress immune responses that are triggered by stress while DHEA enhances or stimulates immune function. Speculation is that in a healthy individual levels of the two hormones might fluctuate with DHEA acting as a check on cortisol so that the immune response is not permanently dampened in response to stress or minor illness and vice versa. Following along these same line, a loss of DHEA production (which occurs in many severe illnesses not just HIV and AIDS) could enhance the suppressive impact of cortisol and perhaps even allow it to continue to dampen immune responsiveness in an unchecked fashion.

A 1993 study from Louisiana State University seems to go a long way toward supporting this theory. Researchers there found a positive link between the immune status of patients with HIV-related illness and DHEA. As the HIV infection became more significant, median DHEA sulfate levels fell and the ratio of cortisol to DHEA sulfate in the blood rose. Researchers theorized that a DHEA deficiency might be causing a worsening of immune function. They arrived at that theory after measuring the levels of DHEA, cortisol, and Helper T-cells in the blood of ninety-eight individuals with HIV. Measures of T-cells are considered a standard method for determining immune status and the progressive destruction of the immune system in people with HIV.

By the end of the study, reserachers concluded that a DHEA deficiency may cause a worsening of immune function. Even though it's not clearly understood, the impact seems so promising that small studies testing DHEA as a treatment for HIV are underway.

DHEA: A buffer against stress?

Serious ills such as cancer and AIDS aside, what kind of impact can DHEA have on more everyday stresses that put wear and tear on the body? Long lines at the bank, a flat tire on the way to work, aggravating co-workers. Do you need DHEA to cope with these more minor insults? That's a valid question. And it's certainly on the minds of researchers as the following new reports on stress indicate.

• Blood samples taken from a group of Ohio State University studies show that levels of natural killer (NK) cells drop during the taking of exams, eventually rebounding to normal levels when stress is resolved. You probably recall that NK are the immune warriors believed to help fend off viral infections and battle tumors.

• A decade of studies from Ohio State University document depressed immune function, delayed wound healing, and increased blood pressure for caregivers of people with Alzheimer's. That's not surprising when you consider that experts suggest that many of these folks are under round-the-clock stress. Even more amazing, these damaging effects last months or years after the spouse with illness dies.

• Researchers at Carnegie Mellon University in Pittsburgh find that major life stressors, such as the loss of a job or a loved one can lower resistance to infection and make a person nearly twice as likely to catch a cold.

While provocative, these findings don't begin to explain the full extent of how different stresses affect immune function. Nor do they explain why some people react negatively to stress and others seem to thrive on it. However, studies show that stress can raise levels of hormones, such as cortisol and adrenaline, that depress the immune system. Maybe the brain is hard-wired to the immune system and hormones act as chemical messengers to convey the message of stress. So the next time you find yourself in a traffic jam or behind the eight ball consider that how you handle this stress could play a role in determining the health of your immune system.

Building the complete immune-boosting lifestyle

As scientists tease out the impact of DHEA and other adrenal hormones on immune function, keep in mind that the current gaps in knowledge don't have to prohibit you from piecing together a program to help keep your immune system robust. Regardless of whether you choose to take DHEA supplements or just to try and boost levels of the hormone naturally, there's a lot you can do right now to boost immune function. Indeed, preliminary findings suggest that what you choose to eat and how your choose to live can have a potent impact on many of the components of internal defense.

- **Throw out the cigarettes**

When healthy nonsmokers and smokers were exposed to a respiratory virus as part of a routine medical study nearly 40 percent of the group came down with a cold, according to researchers at Carnegie Mellon University in Pittsburgh. But smokers were twice as likely to "catch" that cold as nonsmokers.

- **Tune into your spiritual side**

While doing postop follow-up on individuals who recently underwent open heart surgery researchers noticed something unusual about survivors. Individuals who reported finding solace from religious beliefs during their recuperation were three times more likely to survive than those who did not. In an unrelated report, women who underwent hip replacement surgery were able to walk farther and were more upbeat about their situation if they possessed strong religious beliefs. The bottom line: Spiritual beliefs can have a positive impact on recuperative powers.

- **Don't keep emotions bottled up**

When you find yourself replaying a situation over and over in your mind and can't seem to let anger or negative feelings go, consider unleashing the writer within. Scientists at Southern Methodist University in Dallas find that letting go of negative emotions by writing about them in a personal journal may help to boost immune function. Students who spent fifteen to twenty minutes per day, for three or four days, penning thoughts about a traumatic event or an area of conflict in their lives ended up making fewer visits to the doctor in the months following the study than a control group who didn't complete the writing exercise. Five other studies have reached that same conclusion.

- **Don't go overboard on alcohol**

It's long been accepted that continuous heavy drinking can weaken immune response and increase the risk of infection, perhaps even the risk of cancer. But scientists from Lousiana State University wanted to test the impact of social drinking, a few drinks at a party, on immune function. They asked individuals to drink two twelve-ounce cans of beer while eating a pizza (a difficult task but someone has to aid scientists in their search for the truth, right?). A control group did not drink any alcohol. After studying blood samples from the two groups, scientists noticed that the drinkers had lover levels of lymphokine-activated killer cell activity. What does that mean? Speculation is that lower levels of this measure of immune function may equate to a body that is less able to make a clean sweep of foreign invaders—viruses and newly formed cancer cells—and remove these damaging substances from the system.

Researchers aren't concluding that social drinking will wreak havoc on immune function. They simply propose that a few drinks may dampen immune responsiveness to a small degree, much the same way that a viral infection temporarily suppresses immunity. More importantly, they caution people who are taking medications (steroids, drugs to treat diabetes) that suppress immune response to talk with their doctor about social drinking since both the medications and alcohol weaken immune function.

• **Exercise infections away**

In a small study done at Appalachian State University in Boone, North Carolina, researchers found that exercise walkers (people who spent forty-five minutes a day five times per week walking at a moderate pace) suffered 50 percent fewer upper respiratory tract infections than a control group of "couch potatoes." It's far to soon to make predictions about how exercise influences components of the immune system. But it's certain there's no harm in prescribing exercise as a potential therapy to boost immune function.

• **Trim the fat**

Studies clearly show that people who are carrying around too many extra pounds, and an ever-growing number of Americans fit into this category, tend to show subpar results on tests measuring immune response. In addition, experts say that too much fat in the diet—be it from vegetable oils or butter—can dampen immune function. In other words, less is better when it comes to fat.

- **Consider vitamins for vitality**

At Memorial University of Newfoundland in Canada, scientists find that a multivitamin tablet with an additional antioxidant boost (extra amounts of both vitamins E & C) may help keep infections at bay. After following a group of older adults for a period of one year they found that those who took a prescribed daily vitamin with antioxidants were less likely to become sick than adults given a placebo. Interestingly, if these supplement users did succumb to an infection it was usually short-lived and they were plagued with fewer symptoms. People in the supplement group showed increased levels of T-cells and other immune cells that help fight infection.

Following along these same lines, researchers at the USDA Center on Aging at Tufts University in Boston are testing the concept that certain vitamins are capable of rejuvenating an aging immune system. In one of their studies, these scientists noticed that older adults deprived of vitamin B_6 had measurably lower levels of interleukin 2, a T-cell growth factor. In another project, volunteers given daily supplements of vitamin E (800 milligrams) showed improvement in certain key immune responses. Theory has it that vitamin E and other dietary antioxidants (vitamin C, beta-carotene) may help prevent damaging substances called free radicals from causing abnormal changes in cells, changes that may induce heart disease, cataracts, and certain types of cancer.

Q & A

If DHEA and other factors (exercise, diet) improve certain measures of immune response, does that mean I will be at a lower risk for disease if I follow this immune building program?

Unfortunately, that answer is no. Experts say it takes a leap of faith to say that improvements in certain measures of immune response will equate to less illness. Remember, the immune system is a complex network of cells and chemicals that scientists have yet to completely decipher. In addition, you can't ignore the role genetics and other environmental factors play in your risk for illness. Nevertheless, if you employ healthful strategies that contribute to keeping certain cellular responses vigorous, experts do agree that's probably a good thing.

Could DHEA be showing promise as a treatment for AIDS because it boosts testosterone levels, a hormone that is sometimes deficient among people with this illness?

Maybe so. It's difficult to rule out any possibility at this stage of the game since so little is known about HIV and AIDS. Interestingly, researchers at Johns Hopkins Medical Institute report that declining testosterone levels in HIV-positive men may be an early-warning sign of the impending and dangerous drop in weight that often occurs when AIDS begins to develop. Scientists already know that HIV-positive men who lose too much weight before developing AIDS are at risk for earlier death than counterparts who keep weight steady. Knowing beforehand that this weight loss might occur could help in the treatment of AIDS. If testosterone is a valuable marker in regard to this weight issue, perhaps DHEA may also prove to be a kind of diagnostic tool at some point in the future.

Chapter 8

Autoimmune Disorders: Help for When the Body Attacks Itself

An immune system gone awry is a dangerous thing. Instead of distinguishing body organs and tissues from foreign invaders or "nonself" and acting accordingly, the body's defenses launch an attack on "self." These misguided attacks against the body's own cells and tissues can produce a variety of results. Sometimes they "turn on" excessive growth in an organ or tissue. In other cases they cavalierly destroy cells and tissues that perform vital body functions, wreaking havoc with no apparent rhyme or reason. Preliminary evidence suggests DHEA may play a role in treating lupus, an autoimmune disorder that damages the body's connective tissues (skin, joints), blood, and kidneys. However, it's therapeutic role for other autoimmune diseases is less clear.

Some more background on autoimmune disorders

Scientists categorize autoimmune damage as either organ-specific (the damage is confined to one organ) or nonorganic specific (damage is spread throughout the body). Although the underlying mechanisms for both courses of destruction are poorly understood, scientists suspect that immune T-cells may be involved at some point. It's believed that in the everyday production of T-cells the body can make mistakes. That is, make T-cells that don't function properly or that are programmed to recognize "self" or the body's own tissues as a foreign material. Speculation is that the body has a system of recognizing these defective T-cells and destroying them. But some defective T-cells may "slip through the cracks."

It could happen something like this. Scouts on the immune surveillance team roam the body in search of foreign invaders. When they find an enemy substance, immune cells manufacture protein substances called antibodies to protect against that virus or toxic compound that immunologists refer to as an "antigen." In autoimmune disorders the body for some mysterious reason manufactures antibodies in response to its own cells and tissues. These mixed-up antibodies are called "autoantibodies" and can hook up with "self" antigens to form an "immune complex." As these immune complexes accumulate in body tissues they can cause damage, inflammation, and pain. Research shows that nearly every body cell has an antigen called human leukocyte group A (HLA) that helps the body know that this cell is "self." But if these antigens somehow become relabeled as "foreign" the immune system can be fooled into thinking that "self" organs or tissue must be attacked or that "nonself" oganisms present no danger.

If the malfunctioning immune system and autoantibodies are somehow at the root of most autoimmune disorders, you might wonder why researchers aren't testing DHEA as treatment for all of these illnesses. Good question. The best answer is that autoimmune disorders, although they are lumped together under one umbrella classification, progress in different fashions and impact different body systems, making it unlikely that one treatment might work to "cure" or treat all of these ailments. Here's a quick rundown of the influences different autoimmune illnesses have on the body as well as their connection to DHEA.

Diabetes mellitus—Although the designation is considered controversial, many scientists suspect that Type I or insulin-dependent diabetes is probably an autoimmune disorder. Research suggests that the immune system manufactures antibodies (infection-fighting proteins) that mistakenly destroy insulin-producing cells in the pancreas called the islets of Langerhans. People with this type of diabetes must give themselves daily injection(s) of the hormone insulin. However, this is not the case with Type II or noninsulin-dependent diabetes. It is not an autoimmune disorder. (See Chapter 4 for more about this type of diabetes and the impact of DHEA.)

Grave's disease—A few years ago Barbara Bush brought this autoimmune disorder, which is also called thyrotoxicosis, into the spotlight. It results when antibodies inappropriately trigger the thyroid, a small gland at the base of the neck that controls metabolism, to release massive quantities of thyroid hormones. Common symptoms of this "hyperactive" thyroid disorder include: weight loss, rapid heart rate, nervousness, and in some cases bulging eyeballs. Alternative medicine practitioners sometimes use DHEA as an adjunct treatment for this disease but evidence of success is anecdotal. At present, most physicians stick with conventional treatments for this disorder.

Multiple sclerosis—For some inexplicable reason, the body launches an attack on patches of the white coating material (myelin) that surrounds and protects nerve fibers in the brain and spinal cord. The disease can become progressively worse over time and speculation is that an immune system gone haywire may be at fault. Immune cells mistakenly recognize myelin as a "foreign" substance and gradually destroy it, harming underlying nerve fibers in the process. The search for a cure is still underway but researchers say that MS is a very unpredictable disease, one that tends to go its own course. Spontaneous remissions (lasting a day, a month, or even five years) are not uncommon, making it difficult to test potential treatments without a double-blind placebo-controlled clinical trial. Even so, a scientific expert at the Multiple Sclerosis Foundation says there is no strong scientific rationale to even suggest that DHEA could be a potential treatment for MS.

Rheumatoid arthritis—One of the most common and most serious types of joint disorders, rheumatoid arthritis (RA), results when the body's own internal defenses start attacking joint linings. Commonly affected joints include: fingers, wrists, shoulders, knees, hips, and spinal joints in the neck. These joints can become swollen, painful, and stiff, sometimes leading to disability. Treatment may include drugs that dampen immune activity (immunosuppressants), antiinflammatory medications, and drugs that slow progression of the disease. Where does DHEA fit in? That's uncertain. It's interesting that researchers from Spain noticed lower levels of androgens (testosterone and DHEAS) in a group of ninety-nine men with rheumatoid arthritis. And it's interesting that some alternative medicine practitioners report anecdotal success with DHEA as part of the treatment regimen for RA. But so far there are no scientific studies to suggest that DHEA can do anything to help prevent or relieve this potentially crippling autoimmune disorder. In fact, the latest treatment causing excitement among rheumatologists is a class of genetically engineered drugs, medications that are designed to attack the cause of rheumatoid arthritis rather than just the symptoms.

Systemic lupus erythematosus—called "lupus" for short, this autoimmune disorder results when components of the immune system attack the body's own connective tissues and inflammation and pain can be the result. The degree of severity varies from patient to patient. In some cases, damage to the kidney is also apparent. Lupus is ten times more common in women than in men and tends to strike in the second to fourth decade of life. The Lupus Foundation estimates that more people have lupus than AIDS, cerebral palsy, multiple sclerosis, sickle-cell anemia, and cystic fibrosis combined. At last count in 1994 close to two million people in this country had been diagnosed with lupus. Preliminary research suggests that DHEA may be able to help some of the people who suffer with lupus. In fact, the adrenal hormone shows amazing promise as a potential treatment for this baffling disorder.

Why is DHEA being tested as a treatment for lupus?

A number of different drugs have been used in the treatment of lupus. But prednisone, a powerful steroid that turns off the body's production of the adrenal hormone cortisol, is one of the most common. Excitement was high when the drug was first discovered a few decades ago. (It's inventor was awarded a Nobel prize.) Belief was that the medication, because it could eliminate inflammation, might actually cure arthritis and other arthritic disorders. After years of both acute and chronic use, physicians now realize that prednisone's dramatic benefits come with a high price tag. A dose of five to seven and one-half milligrams is generally considered harmless. But larger doses, particularly if taken chronically, can have very, very serious side effects. Some people with lupus don't take any prednisone; others take fifty milligrams per day or more.

Preliminary findings suggest that DHEA may be able to help lessen prednisone requirements in some people suffering with lupus. Before looking at those studies, it helps to understand why both physicians and people with lupus are looking for a way to decrease dependence on steroid medications such as prednisone. Specifically, these are the deleterious side effects both patients and their doctors hope to someday sidestep:

- **Brittle bones:** Let's say you are just starting prednisone therapy. Estimates suggest that you will lose 50 percent of your body calcium supply (most of which is in the bone) in the first six months of treatment. Continue to use the drug and it will leech even more calcium from bone, accelerating your chances of developing osteoporosis. One scientist says that it's possible that a woman diagnosed with lupus in her twenties, after twenty years of heavy prednisone therapy, could have bones similar to that of an eighty-year-old.

- **Unhealthy levels of fat in the blood:** Prednisone alters the balance of lipids (fats) in the blood. It raises total cholesterol levels and lowers HDL, or what is often referred to as the "good" cholesterol. A woman taking large doses of the drug in the years before menopause (when the hormone estrogen normally confers protection against heart disease) may wipe out her natural protection. Heart disease may develop at an earlier age.

- **Less resistance to infection:** Prednisone is a powerful immunosuppressant, a drug that reduces the immune system's ability to fight infection. On the plus side, it helps suppress the autoimmune changes that result in the progression of lupus. The downside is that it also weakens the ability of the immune system to fight other infections be they mild (colds) or major (tuberculosis). Infection is one of the leading causes of death in lupus.

A small pilot study shows promise

Given that one of the main drug therapies for lupus has serious side effects and considering that there hasn't been a new drug to treat lupus for more than thirty years, it makes sense that medical practitioners are shopping around for any new therapy that shows a hint of promise. It also seems logical that the hormone DHEA is on that list of candidates since scientists have long suspected that lupus may be tied in some way to the body's hormonal fluctuations. In animal studies, treatments with DHEA have been shown to cure a lupus-related condition. What is amazing, however, is the magnitude of promise this steroid is showing in the treatment of lupus in humans. It's far from certain that DHEA may be a beneficial drug for people with lupus. But scientists are taking small, careful steps toward building a case for DHEA as just that.

It all started with a tiny 1994 study done at Stanford University in California. In this pilot project researchers wanted only to see if DHEA might have any therapeutic benefits for people with lupus. Women continued with their regular treatment regimens that included prednisone for some and a variety of other drugs. During the study researchers found that they could decrease prednisone dosage as improvements in clinical symptoms among women indicated that less medication was necessary. The participants in the study were a varied group. Some had been diagnosed with lupus only four months before the study started; others had lived with the disorder for five, eight, or as long as ten years. But all of the participants reported feeling better on the DHEA treatment and tests done by their private physicians concurred with those feelings. If the finding that DHEA could help lower steroid requirements could be spread across such a diverse group, wasn't it time to test this astonishing hormonal therapy in clinical trials?

Taking the next research step

The first double-blind placebo-control study with DHEA took place at Stanford University and studied women suffering with mild to moderate cases of lupus. Some of them received 200 milligrams of DHEA for three months or a look-alike placebo. The study found that women receiving DHEA fared better in self-reports as well as on tests of physician assessment, tests that measured specific clinical disease outcomes. At the conclusion of the study researchers offered placebo users DHEA supplements and noticed that they improved in much the same fashion as supplement users in the official study.

Senior author on the study, James L. McGuire, a professor of medicine at Stanford sums up the significance of these findings in a recent news release. "Lupus patients depend heavily on steroids such as prednisone to control their disease. Prednisone, while lifesaving, is responsible for many of the serious complications lupus patients get, including heart attacks, fatal infections, and osteoporosis. Our study shows that patients feel better on DHEA even when they reduce their prednisone dose. It's safety profile appears excellent in our studies, and I am pleased to be involved with the Phase III trials needed to confirm these findings."

Cutting-edge research: Moving into the third phase

After such promising preliminary results with DHEA in the treatment of lupus, Stanford University scientists signed an exclusive license agreement with Genelabs Technologies Inc., a biotechnology company located in Redwood, California, to develop the hormone DHEA as an experimental treatment for lupus. Genelabs filed an Investigational New Drug Application ("IND") with the Food and Drug Administration in the winter of 1993. Gaining an IND allows the company to test the drug as a potential treatment for lupus as well as test its safety and effectiveness.

A just-completed clinical Phase III trial (IND studies progress in phases. The third phase is typically the final one before approval and involves determining the correct dosage as well as safety measures) looked at the issue of whether 100 or 200 milligram doses of GL701 might improve clinical outcome and the degree of symptoms in a larger group of patients with mild to moderate lupus. The study lasted seven to nine months; the length was determined by each patient and the time it took that patient to be able to lower the prednisone dose. Keep in mind that the goal was to see if DHEA could help reduce prednisone requirements. Results will be available sometime in early or mid-1997.

At the same time, Genelabs and Stanford are enrolling volunteers for another Phase III trial. This study began in the spring of 1996 and will enroll 300 patients in about twenty-five different centers around the country.

Lupus Foundation takes a stand on DHEA

In light of these dramatically positive results, people with lupus might want to run right out and order DHEA supplements. The problem, say researchers, is that preliminary studies, no matter how suggestive, are not proof positive of a benefit. They prefer to wait until the results of the trial finished in December (1996) can be tabulated. Until then, no one will really have an answer to the question of whether DHEA can be a valuable treatment for lupus. If you're still tempted to self-treat yourself with DHEA, scientists discourage using DHEA supplements bought over the counter since the amount of DHEA they contain is questionable. They also urge you to seek a doctor's approval and let that physician follow up any treatment regimen. Here is what the Lupus Foundation advises:

> *DHEA is currently being evaluated in lupus patients in controlled clinical trials. Until these studies are completed, we will not know if DHEA is a safe or effective drug for the treatment of lupus. We do not advocate its use, except in these controlled studies where qualified investigators monitor the patients.*

Q & A

If more studies with people are needed to prove that DHEA may be beneficial therapy for a lot of chronic illnesses, why aren't pharmaceutical companies jumping on the bandwagon to study it?

Good question. The pharmaceutical industry is hesitant to fund studies on DHEA for one simple reason: the hormone cannot be patented. DHEA is a natural hormone. If companies sink large sums of money into studying DHEA they won't be able to recoup those investments over the long term since basically everyone already has the formula for this hormone. If no one has the patent, any company could start manufacturing the product once clinical trials prove it useful. In other words, there's no financial incentive to prove DHEA is beneficial.

Aren't allergies another case of the immune system gone berserk? Could DHEA help treat my allergies?

You're right about allergies. They do cause the immune system to overreact to harmless foreign substances (dust, pollen, molds, insect stings, etc.). But these enemies are not quite on an even par with more dangerous foes such as viruses, bacteria, and cancerous cells. When an allergen such as one of the foreign substances above invades the body it is recognized by T-cells. They command B-cells to produce allergy-inducing antibodies. But these antibodies eventually trigger the release of histamine (the substance that produces symptoms such as a runny nose) and so are not as damaging as the autoantibodies that have a hand in autoimmune disorders. Right now it's doubtful that DHEA could play a role in minimizing allergic responses.

Chapter 9

Blocking Tumors, Warding Off Cancer

Despite the fact that cancer is not the leading cause of death in this country (it's second to cardiovascular disease) this is an illness most of us put high on our list of fears. And like researchers, many Americans would like to hear messages about what "cures" or prevents cancer not the 1,001 little things we do that can cause it. Exciting new research suggests that the adrenal hormone DHEA may help fulfill our wish for a prevention strategy. In animal studies, DHEA has been shown to be a potent anticancer agent, blocking the formation of malignant tumors. In addition, scientists have noticed low circulating levels of DHEA and DHEAS among people with breast and prostate cancer. Is there a connection? That's what researchers are trying to figure out. In fact, some of the studies are being sponsored by the National Cancer Institute, a branch of the government's National Institutes of Health.

Battling cancer: Recognizing the enemy

In order to understand what, if any, role DHEA might play in cancer prevention it's necessary to have at least a rudimentary knowledge of how cancer develops. Luckily, advances in molecular genetics, immunology, cell biology, and virology have allowed scientists to gain a clearer understanding of the carcinogenesis process. Not every step in the picture is crystal clear, but researchers do know that when cancer reaches an advanced stage there is less chance for a cure. That's why scientists are searching high and low for effective preventive strategies. Studies are concentrated in two key areas:

Primary prevention—The goal of primary prevention research is to identify and eliminate cancer-causing agents and to locate agents that can block cancer. For example, studies with different populations have given researchers clues about certain environmental factors—smoke, ultraviolet rays, radiation, chemicals—that appear to play a role in causing cancer. The hope is that by systematically isolating agents that cause cancer and agents that may block cancerous growth, researchers will one day be able to put together a comprehensive plan of cancer prevention.

Secondary prevention—The goal of secondary prevention research is to screen individuals who are at increased risk for a particular type of cancer in the hopes that early detection and early treatment will improve survival.

Blocking the formation of cancerous tumors

Some of the most remarkable work in the area of DHEA/cancer prevention comes from the renowned

Fels Institute for Cancer Research and Molecular Biology at Temple University School of Medicine in Philadelphia. Researcher Arthur G. Schwartz and his colleagues have administered DHEA to animals and found that the hormone can inhibit the development of experimental tumors in many tissues—the breast, the lung, the colon, the liver, the skin, and the lymphatic tissue. Unfortunately, most animal studies use extremely large doses of DHEA and skeptics question if these short-term, large-dose studies can apply to people. Nevertheless, it's interesting to see how animal studies suggest an anticancer role for DHEA. Here's a quick rundown of some of those results:

• In a 1995 study published in the *International Journal of Cancer,* researchers reported that after close to four months of DHEA therapy (large doses of the hormone) rats with liver nodules experienced a regression or decrease in the number of nodules.

• Laboratory rats given an injection of DHEA before being exposed to a potent carcinogen that would otherwise promote tumor growth remained tumor-free.

• A report from the Institute of Medical Science at St. Marianna University School of Medicine in Kawasaki, Japan, looked at the ability of DHEA to inhibit the growth of melanoma cells (using melanoma cells from mice) and found that treating cells with the hormone did, indeed, slow their growth.

• Another 1995 report from Japanese researchers suggests that DHEA may help block the progression of radiation-induced tumors of the mammary glands in pregnant mice.

Preventing cancer with drugs: National Cancer Institute studies

Figuring out how to stop cancer in its tracks is an admirable, but so far elusive, goal. At the National Cancer Institute in Bethesda, Maryland, and all across the country at research sites funded by NCI, scientists are trying to tease out the cancer-fighting potential of hundreds of different chemicals and substances. DHEA is on that list. So are a lot of other familiar compounds including nutrients such as vitamin E, calcium, and folic acid and "phytochemicals" (disease-fighting plant compounds) such as genistein, a substance found in soybeans.

It is hoped that by studying these chemicals and figuring out their mechanisms of action, scientists may uncover a substance that can act or help in the development of "chemopreventive" drugs. These chemopreventive medications or substances are defined as agents that will prevent changes that induce or bring about cancer, inhibit it, or delay its progression. Already more than 1,500 naturally occurring and synthetic chemicals have been tested for their "chemopreventive" activity. At this point, most of the work has been done with animals or in the laboratory with human cell cultures. Researchers break these compounds into three basic categories.

1. *Carcinogen blocking agents*—Substances that may block the formation of tumors include nutrients such as calcium; plant chemicals such as polyphenols (fruits, wine) and isothiocyanates. And even drugs such as the nonsteroidal antiinflammatory agents (NSAIDs) used to treat disorders

such as rheumatoid arthritis may offer protection against cancer. DHEA fits into this category.

2. *Antioxidants*—Compounds that neutralize damaging substances called free radicals include nutrients such as vitamin C and vitamin E as well as the vitamin A precursor beta-carotene.

3. *Antiprogression agents*—These substances may stop cancer cells from multiplying or proliferating. On that list are several nutrients such as vitamin E, folic acid, and the trace mineral selenium. A cancer-fighting chemical called genistein has also been isolated in soybeans.

The fact that scientists are pursuing so many different studies puts DHEA research into perspective; as you can see, this hormone is one very small part of the treatments NCI scientists are studying for their anticancer potential. In other words, as compelling as the animal data seem and as suggestive as studies finding low blood levels of DHEA in people suffering from certain types of cancer, researchers are still a long way from proving that DHEA is a solid anticancer-prevention strategy. In fact, when it comes to data from studies on humans much more is known about how antioxidants such as vitamin C and beta-carotene may help ward off cancer.

Claiming the antioxidant advantage

There is some suggestion that DHEA might be an antioxidant with capabilities that are on a par with the nutrient antioxidant trio: vitamin E, vitamin C, and beta-carotene. Antioxidants are valuable chemicals produced by the body naturally (and obtained from certain foods) that are believed capable of neutralizing damaging substances in the body called free radicals. These free radicals form constantly as the body goes about its daily business of functioning; free radicals are waste products in numerous internal chemical reactions. You also absorb free radicals from the environment: cigarette smoke, ozone.

Speculation is that this free radical damage can become more overwhelming to the body as it ages. And instead of keeping up with the repair of free radical damage, the body lets these destructive compounds break and tear DNA (genetic material), allowing changes in cell function that could set the stage for cancer. It may also conspire to alter harmful LDL cholesterol and make it stick more readily to artery walls. Keep in mind that this is all pretty much in the theory stage; many scientists find the concept highly plausible, while others aren't so sure.

Launching your own cancer-prevention program

Obviously, DHEA is far from having a confirmed role in cancer prevention. But don't despair. There are a lot of prevention strategies you can apply that have more positive results. Indeed, estimates are that as much as 80 to 90 percent of all cancer is attributable to environmental factors. (The National Cancer Institute suspects that up to 35 percent of all cancers are diet-related.) Eliminating these risks could prove even more beneficial than DHEA hormone replacement therapy.

Go undercover—Estimates are that roughly 630,000 (out of 700,000) of the skin cancers diagnosed in 1993 could have been prevented if people had slathered on some sunscreen, worn a hat, and otherwise "covered up" to protect skin from sun damage. The most dangerous type of skin cancer, malignant melanoma, is on the rise in this country.

Shun tobacco—Not surprisingly, cigarette smoking is responsible for more than eight out of ten cases of lung cancer and approximately 30 percent of all deaths due to cancer. The numbers are astounding. For instance, in 1993 the American Cancer Society predicted that approximately 160,000 deaths from cancer would transpire solely because of tobacco use.

Eat less fat, more fiber—That what you eat is related to cancer risk is a given. Experts haven't been able to nail down a specific anticancer eating regimen. But they do know that excess fat—lots of rich pastries, chocolate bonbons, and hefty steaks (somebody has to be eating all this red meat if the burgeoning resurgence of steakhouse restaurants is any indication) is not good. Conversely, diets high in fiber—whole grains, fruits, and vegetables—seem to confer a lower cancer risk.

Stocking your pantry with a cancer-fighting arsenal

As mentioned above, a diet that is low in fat and rich in fruits, vegetables, and whole grains is your best primary food weapon in the fight against cancer. (See the DHEA Food Program in Appendix I.) But don't let your program stop there. Preliminary research continues to uncover a wealth of potent cancer-fighting chemicals in some of the most unlikely pantry staples. Here's some of the latest news on this "phytochemical" forefront. Specific anticarcinogens that have been identified in foods include nutrients like vitamins A, C, and E and the vitamin A precursor beta-carotene.

- **Broccoli**—Researchers from Johns Hopkins Medical Institute in Baltimore were the first to discover an anticancer agent in broccoli called sulphoraphane. In laboratory studies this chemical helped boost the effectiveness of enzymes that block cancer. In animal studies, sulphorophane is able to block mammary tumors.

- **Garlic**—Several researchers have isolated chemicals in garlic that appear capable of halting the formation of carcinogens (cancer-causing compounds) by indirectly activating the enzymes that neutralize these compounds. Researchers conducting the Iowa Women's Health Study have discovered a link between diets rich in garlic and a lower colon cancer risk.

- **Soybeans**—Chemicals found in soybeans called isoflavones may help encourage enzymes that fight cancer, particularly cancer of the breast. Soybeans are turning up in more places in the supermarket: veggie burgers, snacks, and, of course, tofu. (See Chapter 12 for more about the health benefits of soy as a replacement for estrogen.)

- **Tea**—Preliminary studies suggest that polyphenols, antioxidant compounds commonly found in the green teas sold in Asian markets and some supermarkets, may help prevent esophageal cancer.

Chapter 10

Hormone Replacement Therapy: Will It Be Testosterone, Estrogen, DHEA, or All Three?

When the topic of hormone replacement comes up, most of us immediately think hot flashes, estrogen, and menopause. But new research seems to indicate that there's benefit to be had from some of the other sex hormones, namely DHEA and testosterone. Speculation is that restoring dwindling hormone supplies to youthful peak levels, particularly the androgens and sex hormones, may return everything from sex drive to immune function to weight levels to those you experienced while in your prime.

With that kind of sales pitch, who would think twice about hormone replacement therapy. Still, even with treatments that have been around for several decades, such as estrogen replacement therapy (ERT), it's still important to weigh the potential benefits with the risks. Before we look at using DHEA to replace hormone levels, let's look at a couple of hormone therapies already in use.

Replacing estrogen after menopause

In the early years before menopause, the hormone estrogen gives women a distinct health advantage. It probably helps keep cholesterol levels low and favors the storage of fat in a more benign area, the hips and thighs. But as women get older estrogen levels decline and levels of adrenal steroids increase. Since androgens favor the storage of fat in the abdominal area (the "apple" body shape) women may start developing fat around the belly.

Studies show that estrogen replacement therapy (ERT) can help to reverse these unhealthy changes and thereby lessen a woman's risk for heart disease. Mayo Clinic researchers estimate that if all the women aged forty to fifty-nine were given ERT, risk for heart attacks could drop as much as 45 percent. ERT also appears to help keep bone dense and thereby lessen the risk for the bone-crippling disease osteoporosis. Are these benefits worth it? That's something you'll want to discuss with your physician. It's important to make this decision based on your individual health profile.

If you are leery of ERT you might want to think about the results from a recent yet-to-be published study done at the University of Illinois at Urbana-Champaign. Scientists there find that foods containing soy protein may be a substitute for estrogen. Specifically, they fed a group of postmenopausal women diets that contained forty grams of isolated soy protein (in baked goods and beverages) as part of a low-fat low-cholesterol diet. LDL or the "bad" cholesterol dropped about 8 percent; HDL or the "good" cholesterol jumped more than 4 percent. The soy protein also increased bone-mineral density.

Weighing the risks and benefits

Estimates are that estrogen replacement therapy is safe and tolerated very well by nine out of ten postmenopausal women. But there are exceptions. A small percentage of women experience intolerable side effects such as erratic bleeding and spotting, breast tenderness, and excruciating headaches. Others find that their overall health and the presence of other illnesses such as acute liver disease or active breast cancer absolutely contraindicate the use of estrogen. But with all the evidence supporting a protective role for estrogen against both heart disease and osteoporosis, many of these "absolute" contraindications have become "relative," argues Dr. Karen Miller, M.D., of the University of Utah Medical Center's Department of Obstetrics and Gynecology.

In other words, in light of the potential stellar health benefits of estrogen replacement many postmenopausal women seem likely candidates for this type of replacement therapy. If you are trying to make a decision about ERT you might want to talk over Miller's list of contraindications with your doctor.

Could it be "male menopause?"

It's not only women who experience a decline in hormone levels with age; men produce less of one key sex hormone as they grow older. After a man reaches the age of forty-five to fifty, levels of the male hormone testosterone begin to decline. However, most men don't seek medical help for the "change," even when these lower levels result in symptoms such as impotence or loss of sexual desire. It's not male menopause so to speak, or is it? Estimates are that somewhere in the neighborhood of four to five million men have a testosterone deficiency but only about 5 percent seek treatment. Many experts think that now there are better methods of replacing testosterone, the numbers of men who seek treatment for a testosterone deficiency could dramatically increase.

New studies suggesting a whole host of health benefits for testosterone therapy could also increase interest in this potent male hormone. At a recent annual International Congress of Endocrinology meeting researchers from Johns Hopkins Medical Institute reported that testosterone affects learning ability. Men with low testosterone levels were placed "on" and "off" testosterone replacement therapy and given an assortment of learning tests. Scores were higher on tests measuring visual and spatial skills. Scientists also are studying the connection between declining testosterone levels in men with AIDS and weight loss.

Which type of supplemental testosterone?

Doctors have been using testosterone replacement therapy on a small number of men for years. Some say that the unpleasantness of replacement therapy probably helped to turn many men who could benefit from testosterone away. With improvements in hormone delivery methods (patch, pill) expectations are testosterone replacement may become just as common in men as ERT is with women. If you're a man and find yourself grappling with the decision of whether or not to replace waning levels of testosterone, consider this expanded list of treatment options.

Testosterone Injections: For years this method was the only option for most men. It consisted of a one-time injection of testosterone every two to four weeks. Men on the receiving end of the treatment often complained of discomfort comparing the injection (which has to be placed in deep muscle) to "a shot of hot lead in the buttocks." Another downside of this delivery method is the extreme fluctuations in hormone levels at the outset and near the end of treatment. Levels of testosterone are high after the injection and then dwindle in the days before the next injection, not a smooth release like the patch.

Patch therapy: Early testosterone drug patches were applied to the shaved scrotum. But in the fall of 1995, the FDA approved a small round skin patch that could be placed on any fleshy part of the body including the arms, legs, or back. Called Androderm (made by Theratech and marketed by Smith Kline Beecham), it delivers a time-released dose of testosterone that is meant to closely mimic the body's naturally fluctuating hormone levels. Usually men are instructed to wear two patches—each patch contains five mg of testosterone, over a twenty-four-hour period. *Note:* The patch is approved for men only; levels are too high to be considered safe for women.

Pill/Tablet: Testosterone in tablet form is available in many European countries but has not been approved for use in the United States. But that situation may be changing. Researchers at Johns Hopkins Medical Institute have been testing a new tablet form of testosterone on men with low testosterone levels with good results. In a recent study they gave thirteen men either a testosterone tablet that could be dissolved in the mouth or a look-alike placebo. After three months the men receiving the active testosterone treatment reported normal sexual function. That leads researchers to conclude that tablets are as effective as testosterone injections but without the wide fluctuation in hormone levels and without negative side effects such as mood swings.

Can women benefit from testosterone replacement?

It's well known that women also produce testosterone, but in much smaller quantities than men. In the past research has focused more on how excess levels of testosterone might harm a woman—androgenization, the development of male secondary sex characteristics, an unhealthy lipid profile, etc. But one physician contends that this focus on excess testosterone overshadows the fact that some women may actually have a testosterone deficiency, a deficiency that could influence libido and energy levels in the postmenopausal years. Very little research has been done recently in this area. However, if you're wondering if your testosterone levels are low, you could have a saliva assay done. (See Chapter 11.) You might also want to pick up a copy of the book: *The Hormone of Desire: The Truth about Sexuality, Menopause and Testosterone* by Boston psychiatrist Susan Rako (Harmony Books, 1996). Rako has looked into this area quite extensively and offers some insights that your doctor probably isn't even familiar with since research in this area is definitely scant.

Is America ready for DRT: DHEA Replacement Therapy?

Although studies confirm that blood levels of DHEA dwindle with age, there is proof that the amount of variance within a particular age group can be quite wide. Before you consider replacement therapy see what researchers know about how these three important factors can influence blood levels of DHEA:

- **Genetics**—When you factor out age, studies show that there is remarkable similarity in DHEA levels within a family, leading many to believe that there could be a genetic component that directs some of the decline in DHEA that seems to occur with age. Some scientists point out that the hormonal milieu in a healthy older person can often closely mimic that of a younger person. In other words, there's no telling what your DHEA level is without checking it. Don't simply assume that your DHEA is low because several of your fifty-year-old friends have low DHEA levels.

- **Physical condition**—Research indicates that DHEA levels tend to be lower among people who are ill (regardless of age). A German study finds that DHEA levels drop after surgery. But scientists point out that low DHEA levels during illness may be a temporary side effect of illness not necessarily a DHEA deficiency. Indeed, the body's reaction to any kind of stressor (and illness and surgery definitely fit into this category) is the same. Subjected to stress it responds by telling the adrenals to produce more cortisol and less DHEA. Keep in mind that if you aren't taking care of your health—getting plenty of sleep, eating healthfully, exercising—simply improving your lifestyle habits might be enough to help raise your DHEA level.

- **Age**—Blood levels of DHEA dwindle with age. Estimates are that, starting at about age twenty-five or thirty, the body slows production of DHEA so that blood levels gradually begin to drop. By about age seventy you are producing only about 10 to 20 percent of the DHEA manufactured in youth. Most of the data looking at how DHEA levels relate to age is cross-sectional. That is, it looks at a wide cross section of the population at one particular point in time. Some scientists would like to see longitudinal data, reports where the same individuals are followed over a period of years. No doubt DHEA levels will still show a decline but researchers may be able to more carefully characterize that decline in different segments of the population and perhaps figure out who among us might benefit from DHEA replacement therapy.

Considering the conversion factor

All the hype about what DHEA might do in animals tends to cloud the issue of what it might do for your individual health. Scientists point out that the transformation of adrenal precursor steroids such as DHEAS and DHEA into androgens and/or estrogens may depend on the availability of certain enzymes that initiate and direct the conversion process. A brand-new area of endocrinologic research is focusing on reactions that occur inside the cells of tissues throughout the body. The science is called intracrinology and researchers are making rapid progress identifying the enzymes that may influence DHEA levels as well as the genes that direct them. Indeed, many researchers would feel more comfortable if the underlying mechanism by which DHEA works could be outlined.

DHEA supplements not without risk

Researchers say there is a long history of methyltestos-terone (synthetic testosterone) use in the '60s and '70s for men with aplastic anemia. Initially this replacement therapy seemed beneficial but down the road problems with liver damage and liver cancer appeared. Many scientists worry that DHEA, another powerful steroid hormone, might be headed down the same path. They caution people who are self-administering this hormone that it is not an innocuous substance. That is, just because it's natural doesn't mean its nontoxic. Here are some of the risks you need to be aware of if you're considering taking DHEA supplements:

- **Liver damage, liver cancer**—Animals studies show that DHEA is a potent hepatocarcinogen. Simply put, it causes liver cancer. Baylor College of Medicine's Dr. Peter Casson reports that one female study participant taking 150 milligrams of DHEA showed signs of jaundice and other symptoms indicative of liver trouble after only a brief period of supplementation. Casson says he reported the finding to the FDA and now does periodic liver function tests on participants in clinical trials.

  Why should the liver be harmed by levels of a hormone that occurred naturally in youth? And is oral supplementation really natural? It doesn't mimic the circadian fluctuations in hormone levels that occur naturally in the body. Perhaps in the future scientists may be able to develop safer administration methods for DHEA so that it doesn't tax the liver. In the meantime, it's important that you seek a doctor's guidance if you plan to use DHEA.

- **Prostate enlargement**—Preliminary evidence suggests that men may experience an enlargement of the prostate gland (hypertrophy) with continued DHEA supplement use. Some scientists suspect prostate cancer, which is believed to be hormone-driven, is another risk. The Life Extension Foundation, one of the mail-order companies selling DHEA supplements actually warns consumers about this risk. Their catalog states: "Men with prostate cancer and women with reproductive cancers should avoid DHEA. Before taking DHEA you should get a DHEA sulfate test. Men should also get a PSA test and a PAP test. Women should get a C.A.-125 (ovarian cancer) test."

- **Virilization**—In the Drs. Morales, Nelson, Nolan, and Yen study of DHEA supplementation in older men and women at the University of California at San Diego, one woman developed facial hair while taking a 100-milligram dose of DHEA over the course of six months. Facial hair disappeared when DHEA supplements were discontinued. In another study with postmenopausal women, mild acne (which was treatable) plagued several participants. An unrelated report from the University of Tennessee finds "supraphysiologic" (abnormally high) androgen levels in women given fifty milligrams of DHEA for only three weeks. Finally, the FDA has received a report of virilization (the development of male secondary sex characteristics in a woman) on a dose of 200 milligrams of DHEA. Many researchers feel that reducing the dose to twenty-five milligrams or less will help avoid excessive androgen production in women. It may also turn out that different methods of administering the hormone (transdermal, non-oral) could enable DHEA supplements to bypass the liver where much of the conversion to potent androgens occurs.

- **Unhealthy alterations in blood lipids**—It doesn't seem to happen in men, but in preliminary findings with women some scientists see a disturbing trend. DHEA supplementation seems to markedly and adversely affect the level of some fats in the blood. It's too soon to report any firm data. But women should keep an eye on this issue. One researcher, if he can't convince his female patients to stop taking DHEA, does periodic blood lipid profiles to make sure that DHEA isn't negatively influencing lipid levels.

The great hormone debate

For years, physicians who prescribe alternative medicine therapies have been using DHEA to treat everything from chronic fatigue syndrome, to arthritis to a waning libido. Although there is a lot of anecdotal evidence there haven't been any scientific studies to support these claims. Chances are you may already know someone who is taking DHEA. Here, in a nutshell, are the two general lines of opinion on the subject of the use of DHEA supplements.

POINT

Self-supplementation is too risky.

Many experts feel there is far too little research evidence to support using DHEA outside of a clinical trial. They point out that those promoting DHEA as an "antiaging" elixir are ignoring other factors—diet, environment, lifestyle—that can influence the aging process and the progression of chronic illnesses such as heart disease, cancer, and obesity. Putting too much emphasis on one magic bullet, DHEA could have a disastrous backlash. Moreover, do we really know enough about the safety issues? To be fair, many of the scientists in this group are broadminded about DHEA's potential and even excited. But they think research should be carried out in a careful clinical fashion, not willy-nilly with people popping DHEA pills (with who knows how much of the hormone inside) as desired.

COUNTERPOINT

Replacing hormones is not unnatural, nor is it risky if a physician supervises the therapy.

A smaller but much more vocal group of alternative medical practitioners have been prescribing DHEA, and testing it on themselves for years without any reported problems. So have some of the big names (the older researchers in their seventies and eighties) in DHEA research. They point out that hormone replacement therapy is nothing new; estrogen replacement has been studied for several decades. Their contention is that if DHEA supplements are taken under the guidance of a physician, and in doses that replace the levels of the hormone found naturally in youth, there shouldn't be any problem.

YOUR DECISION

This is medicine in the nineties. Chances are you'll hear more details than you want to about DHEA from friends, family, coworkers, and even the media since health care is now played out on television, in newspapers, and in popular magazines. But the final decision is up to you. And that's good. Taking responsibility for your health is important. Just keep in mind that unlike melatonin, DHEA is not a harmless substance. It's a powerful hormone and use of it is something that common sense dictates you discuss with your doctor.

Q & A

Is DHEA an immune booster?

DHEA probably does have an immune-enhancing effect, although it appears to be on only certain components of the immune system. Aside from the compelling animal data, there are several in vivo (studies on human immune cells in the lab) that show DHEA can stimulate immune cells. A clinical trial on postmenopausal women that found three weeks of DHEA supplementation (fifty milligrams) lead to increased numbers of Natural Killer (NK) cells, immune cells that ferret out cancerous cell growth and viral infection. Belief is that any immune-enhancing effect attributed to DHEA holds true for both men and women.

Does it protect the heart?

If DHEA is cardioprotective, the effects appear confined to men only and are minimal. One human clinical trial found that DHEA supplements (1,600 milligrams a day for twenty-eight days) given to a handful of healthy young men did help lower LDL, or the so-called "bad" cholesterol. But long-term safety of these kinds of massive doses are questionable, particularly when there are so many other safe and effective ways to lower LDL such as diet, exercise, or lipid-lowering medications. And one population study did link low levels of DHEA to increased risk for heart attack. But others have not. In other words, there is not enough data to support any firm conclusion about DHEA as a heart-protective agent that may explain why the American Heart Association has been conspicuously silent on the subject. DHEA appears to have no cardioprotective impact in women and preliminary findings suggest it may adversely alter blood lipids. Put another way, if you are a woman DHEA may be more harmful to the heart than helpful.

Will DHEA battle the bulge?

Unfortunately for women, DHEA appears only to exert its antiobesity effect on men, and a pretty mild effect at that. Several studies find that men taking DHEA supplements gain muscle mass and lose fat over the course of a few months. However, this is no miracle diet drug. Most experts report only a mild change; roughly a 5 percent drop in body fat stores seems to be the average. While shedding any amount of body fat is laudable if you are overweight, a 5 percent change is not going to make you svelte overnight. Exercise and a low-fat, high-fiber style of eating can probably help you lose that much fat in half the time.

Is this hormone an anticancer agent?

As you no doubt realize, cancer is a complex disease for scientists to study. It has a long latent period (cancer can take decades to develop) and seems to be influenced by numerous factors including genetics and the environment. The National Cancer Institute is studying DHEA as a potential chemopreventive agent, one that may block the formation of malignant tumors, but there are hundreds of other anticancer agents on the same list. In other words, don't expect DHEA supplements to ward off cancer. Don't even expect quick answers to the cancer question; many scientists wonder if it is even feasible to do clinical trials to test DHEA's anticancer potential since these gigantic undertakings are so expensive and many other substances appear more promising. It's entirely possible that DHEA's immune-enhancing ability might help to explain any anti-cancer effects seen in animal studies. Expect scientists to explore this immune connection first.

Can DHEA elevate mood and increase energy?

It's difficult to ignore the nicely done double-blind placebo-control DHEA supplement study (on humans not animals) of Samuel Yen and colleagues at the University of California at San Diego. These scientists find that older men and women given fifty milligrams of DHEA for three months report improved feelings of "well-being." That is, increased energy and a better ability to handle stress. Some flies in the ointment: Yen and his colleagues didn't measure the "well-being" factor in a longer follow-up clinical trial of DHEA supplementation that lasted six months. And another thing, how does one measure "well-being" scientifically? That wasn't explained in the published study that made some researchers wonder. Another puzzling finding: a group of older women in another double-blind placebo-control trial (measuring immune response to DHEA) never guessed which treatment they were on. If DHEA's mood-elevating potential is so dramatic scientists wonder why these women never noticed any difference while on placebo and while on DHEA. This one's too close to call. If DHEA has any impact on mood, it's likely to be mild.

Is DHEA an aphrodisiac?

While this is a big reason why many alternative medicine practitioners prescribe DHEA supplements, so far proof that DHEA revs up the sex drive is purely anecdotal. Chances are you might have a friend, spouse, or family member who is using DHEA for this reason and claims it's a rousing success. Interestingly, in the carefully controlled double-blind placebo-control study by Samuel Yen and colleagues, there was no difference in libido for placebo versus DHEA supplement. In other words, when people didn't know what kind of treatment they were getting—placebo or hormone—libido remained unchanged. Could it be that when people think DHEA is going to increase libido it does? Some scientists say the brain is perhaps one of the most powerful sexual organs.

Chapter 11

Everything You Need to Know About DHEA Supplements: How Much, What Kind, Where to Buy?

Given the limited amount of research to date on DHEA, you'll want to take care in planning a DHEA supplement regimen. There are a number of factors to consider, not the least of which is finding a physician to help set you on a safe course since the side effects of inappropriate DHEA supplementation can be quite serious.

DHEA: Figuring out how much to take

As with any available over-the-counter supplement, you are in the driver's seat when it comes to determining dose. The problem is that at this point in time doctors, even the few who take DHEA supplements themselves, say that there is no one universal dose of DHEA that is good for everyone. Indeed, most researchers suspect that the dose that might benefit men is different from that for women. Baylor College of

Medicine's Dr. Peter Casson told colleagues at a recent medical meeting that he doesn't prescribe DHEA to his patients and, in fact, tries to discourage them from using the hormone until more research is available on the risks and benefits. However, in clinical studies Casson says that "it appears the appropriate dose of DHEA is twenty-five to fifty milligrams per day in women, depending on their endogenous DHEA-S levels." And twenty-five milligrams is probably closer to the mark for most women. Researchers using DHEA and alternative medicine practitioners usually prescribe fifty milligrams per day for men. Keep in mind that there have not been any studies to determine effective dose, timing of administration, or duration of DHEA use. Nothing is known about the long-term side effects of replacing this hormone.

However, among the handful of researchers who are conducting DHEA research and taking the supplement themselves, speculation is that it's probably safe to take replacement levels of the hormone. That is, enough of the hormone so that blood levels of DHEA return to youthful levels. Of course, in order to do this you must have some idea of the current level of DHEA in the blood.

Is your DHEA level low or high or somewhere in between?

The best way to find out DHEA levels is to have your doctor take a blood or saliva sample. Both can be sent to any number of labs across the country that do assays of hormone levels. Either method is highly accurate and your doctor can help you evaluate and interpret the results. Although most physicians don't recommend having hormone levels tested without a doctor's supervision, if you're simply curious about where you stand in respect to DHEA or other hormones, there are a few companies, including the two listed below, that will send you a do-it-yourself saliva test kit. You'll have to send a saliva sample to their lab for testing. But the process is fairly simple. You spit some saliva into a plastic test tube and ship it off to be analyzed. Kits come with instructions about when to take the sample (usually in the early morning for most hormones when production is at its peak).

Aeron Lifecycles
1933 Davis Street, Suite # 310
San Leandro, CA 94577
1-800-631-7900

Kits are sent via regular mail, although if you're in a hurry two-day air service is available. A prepaid shipping container is provided for return mail since, according to the lab, hormones are stable at room temperature. Some of the hormones the company will measure include: estrogen, testosterone, DHEA, and progesterone. Prices may change but at the time of publication a one-hormone assay cost slightly less than $50. Having several different hormones assayed lowers cost progressively with a four-hormone assay costing just under $150. Shipping costs are an extra $5 per kit. California residents will be charged a sales tax.

National Biotech Laboratory
13758 Lake City Way, NE
Seattle, WA 98125
1-800-846-6285

Kits are sent via regular mail with a requisition form to be filled out by your doctor. However, you don't need the requisition form to receive test results. After you return the kit with a saliva sample, results will be shipped to your home address along with a listing of "normal" values or ranges for each specific hormone tested. Prominently displayed on that list is a disclaimer that the test is neither a diagnostic or treatment tool. While the company prefers not to disclose fees for publication, they will quote prices to consumers over the phone.

Don't expect an interpretation of results, however. In fact, order takers will encourage you to talk with your doctor about the results. One warning: don't floss or brush teeth before producing the sample as that could contaminate saliva with blood particles and alter test results.

Just how accurate are saliva tests?

David O. Quissel, at the University of Colorado School of Dentistry in Denver, reports that standard plasma (blood) sampling techniques or urinalyses do not provide the optimal sampling conditions needed for assessing endocrine (hormone secretion) function. Nor are they the best strategy for monitoring plasma steroid levels during medical studies, which is why saliva assays are being used more and more often. The advantages of saliva hormone assays include:

- The procedure is noninvasive. No needles. No blood.
- Samples are easy to collect, store, and send off for analysis.
- The analysis procedure is automated; less chance of human error.
- Saliva hormone levels accurately reflect plasma hormone levels.

Tablet Talk

If you find yourself supplement shopping you'll want to get up to speed on the latest lingo. Being familiar with terminology allows you to get the best deal for your money. And believe it or not, there are some big differences among formulations. Understanding these three key terms (many of the phone operators that sell DHEA don't even know what some of these monikers mean) can convert you into a savvy shopper instantaneously.

• **Sublingual**—tablets that are dissolved under the tongue. This formulation allows for a hormone or other medication to be absorbed directly into the small capillaries under the tongue. Do sublingual tablets have any advantage over oral tablets? Researchers at the University of California at San Diego tested both selections to determine if they might have a different impact on blood levels of DHEA. The conclusion: Both choices appear about the same over the long term. But the sublingual route of administration did produce a more marked early rise in DHEA levels shortly after it was administered. Incidentally, these scientists chose oral DHEA tablets for their clinical studies because of "ease and reliability" of administration.

• **Micronized**—pulverized into minute particles. Researchers typically use micronized DHEA (particles are ten micrometers in diameter or less) in clinical trials instead of crystalline DHEA because it may partially bypass liver transformation into more potent androgens. Predictions are that some day DHEA may be able to be given via a nonoral route—under the skin, in the vein, transvaginally—to avoid hepatic (liver) androgen effects. (Remember, DHEA in extremely high doses is a liver carcinogen in laboratory rats. Being able to avoid excess encounters with the liver route and travel directly to tissues may hold a distinct advantage.)

• **Pharmaceutical grade**—Researchers often talk of a pharmaceutically pure hormone product when discussing the potential benefits of DHEA. That's what they purchase directly from pharmaceutical companies for clinical trials. This assures that they are getting both a pure substance and the dose they need to test. (Any Investigational New Drug (IND) that researchers purchase for study is usually accompanied by documentation of purity.) Some health food store supplements are catching on that shoppers know about this terminology and so are using terms such as "pharmaceutically pure" to make a product sound inviting. It's doubtful that this terminology means anything. Your best bet for pharmaceutical-grade DHEA are compounding pharmacies that purchase their raw materials from a major pharmaceutical company.

Purchasing Pointers

Although supplements of the hormone are just making their way into major discount chains and most neighborhood pharmacies, health food stores carry a wide variety of DHEA supplement products. And mail-order companies and compounding pharmacies, some of which advertise DHEA prominently on the Internet, are another source.

Keep in mind that over-the-counter food supplements are not monitored by the government for purity, biological availability, effectiveness, or safety. In other words, you're on your own when it comes to quality.

Are you getting what you pay for? That's a good question. One scientist suspects, at their worst, what is sold in most health food stores is probably just "talc." Essentially these pills may contain only enough biologically active DHEA to affect a small mouse.

Compounding pharmacies that get their DHEA from major pharmaceutical companies, are likely to be more reliable sources of the hormone.

Compounding pharmacies make each single prescription—dosage, form of medication (tablet, cream, capsule)—from "scratch" using the raw materials. Those raw materials, mostly natural or synthetic hormones, often come directly from major pharmaceutical companies. Compounders then use their own tablet presses to make tablets, capsule machines to make capsules, etc. They also make up prescriptions for creams or gels in any dosage. If you're allergic to lactose (a binding material used in some pills) or a coloring dye, these products can be left out of a prescription. Or if a doctor wants to fine-tune your estrogen replacement therapy to include a variety of natural hormones in differing amounts, the compounding pharmacist can make up a single prescription tailor-made to your needs. At one time, the majority of pharmacists in this country were in the practice of both "compounding" individual prescriptions and dispensing ready-made drug tablets. Today, your typical neighborhood pharmacy is mainly in the practice of dispensing ready-made drugs.

Still, this is a situation where the buyer has to beware. It's probably best to choose your supplement from large, well-known pharmacies as they have the most at stake when it comes to safety and reliability. It doesn't hurt either to question your source about quality control, supplement ingredients, etc. In addition, there are two important pitfalls you'll want to watch out for before making your purchase:

DHEA

1. Be careful of the advice given by clerks in health food stores and over-the-phone order takers. They are salespeople, not medical experts. The best person to answer your questions about developing a DHEA supplement routine is a physician. He or she can monitor your blood levels of DHEA and do other tests (lipid profile, liver function, prostate specific antigen PSA for men) to make sure the dosage you take is safe.

2. Don't purchase supplements that say their DHEA comes in the form of wild yam extracts. One site on the Internet claims that exotic varieties of discorea (wild yam) *"serve as a precursor for the body's endocrine system to manufacture DHEA."* Don't count on it! Despite what label claims may say, the body cannot convert plant steroids (from wild yams or other plants) into DHEA.

International Academy of Compounding Pharmacists

1-800-927-4227

The IACP has over 900 members located in every state, making the list of locations far too numerous to print. But if you call the toll-free number above expect a response within the hour from IACP's Shelly Schluter. She'll plug your zip code into IACP's computer mapping system and it will flag the closest compounding pharmacy location in your zip code, one she says is probably literally right down the street. Often Schluter has a pharmacist at a nearby location contact you directly. *Note:* Although some of the pharmacies do mail prescriptions, Schluter says compounding pharmacists typically prefer personal contact. They like to develop an ongoing relationship that includes both patient and physician.

It's probably best if you resist the temptation of grabbing DHEA supplements from the shelves of your health food store or supermarket. While this may seem like the likeliest and most convenient way to shop, health food shops and supermarkets are questionable sources of the hormone. Again, nearly every scientist doing DHEA research makes a special point to mention that there is no way to know how much DHEA, if any, is in over-the-counter products. Remember, these are dietary supplements so Uncle Sam is not watching what goes into them.

Mail Order Sources

Wherever you decide to look for DHEA keep in mind that the FDA is not monitoring any of the finished products. You already are aware that it's unlikely supplements found in health food stores contain much DHEA. But no one has done tests on products that come from mail-order or compounding pharmacy sources. Granted the raw materials and production methods of hormones are monitored by FDA since these products are made by major pharmaceutical companies. But FDA doesn't follow the path any further. That means you'll need to assess the reputation of any company or pharmacy selling supplements. Indeed, many research scientists say that the only way to be sure what you are getting is DHEA is to hire a lab to perform a chemical analysis. That's obviously not realistic. But keep in mind that as the buyer you will need to scrutinize the supplier. Ask questions and if you are not satisfied with the answers consider looking for another source. With that in mind, here is a partial list of some of the companies that ship DHEA through regular U.S. Postal channels. After each listing are some details provided by the companies.

Bajamar Women's Heathcare Pharmacy
9609 Dielman Rock Island
St. Louis, MO 63132
1-800-255-8025
(314) 997-2948 for FAX orders

Bajamar is a large, well-known compounding pharmacy that operates in St. Louis and can be accessed via the Internet, by phone, or by mail. If you wish to place a mail order, you'll need to write or call for a pink mail order form. Any nonprescription items—DHEA, nutritional supplements, books, and tapes—can be ordered by mail. Hormones such as estrogen and testosterone require a doctor's written prescription. The United Parcel Service (UPS) is the standard method of delivery. UPS third, second, and next-day air delivery are available for an extra charge. You can pay for your order by check, money order, credit card (MasterCard, VISA, or Discover), or COD.

For Your Health
13758 Lake City Way, NE
Seattle, WA 98125
1-800-456-HEAL

This mail-order retail outfit is located in the same building as National Biotech Laboratories, the company mentioned above as a source of saliva assay testing. The two are affiliated and it's this division that markets a wide variety of dietary supplements. The company sells DHEA in five to fifty milligram doses and says it is one of the few places consumers can purchase sublingual tablets of the hormone. A free quarterly newsletter about current supplement concerns (including DHEA) is provided to customers. Delivery is by regular mail; credit cards and checks are accepted for payment.

Life Extension Foundation
995 Southwest 24th Street
Fort Lauderdale, FL 33022
1-800-841-5433

The Life Extension store has been selling vitamins, nutrients, and other supplement products for more than a decade. They are not a compounding pharmacy but ship pharmaceutical-grade DHEA, micronized is one option, all over the world. A fact sheet produced by the group reports that the Foundation has been "in a state of war since 1985" with the FDA over access to dietary supplements. In the 1996 directory of Life Extension nutrients and drugs a passage on DHEA reports a summary of the latest research findings. Anyone can order DHEA from the foundation but Life Extension offers a 25 percent discount on all supplement products to members. Membership costs $75 per year (you can put it on your credit card) and includes a subscription to the foundation magazine as well as resource information about alternative medical treatments and doctors who prescribe them. Checks or money orders are also accepted and mail-order requests can be sent to Box 229120, Hollywood, FL 33022.

Supplement shopping on the Internet

If you're hooked up to the Internet, or if your local library has access to the World Wide Web, you can shop for DHEA supplements via computer. A partial list of options is outlined below. Or if you have the time, access one of the major search engines (Alta

Vista, Yahoo, Webcrawler, etc.) Then plug in the term "DHEA" and watch the computer light up with thousands of different sites on the Internet where this hormone is mentioned. A lot of the sites are repetitive; some companies are into hawking their products in every conceivable "nook and cranny" of cyberspace.

Bajamar Women's Healthcare Pharmacy

Internet address: http://walden.mo.net/~bmizes/hormones.html

This St. Louis-based compounding pharmacy offers on-line reading materials about natural hormones. It's user-friendly and breaks down information into easy-to-access categories with separate sections on the more popular replacement hormones estrogen, DHEA, and testosterone. You can download information (newspaper articles, medical studies, in-house publications) or have it sent to you E-mail. And while they can't give advice, there is access to an "Ask the Pharmacist" service on line. Or you can talk with a pharmacist by calling 1-800-255-8025.

Tenzing Momo's DHEA page

Internet address: http://www.tenzing.com/dl.html

This cyberspace site bills itself as one of the "oldest herb stores in the U.S.," and "the lowest-priced DHEA on the net." It promotes pharmaceutical-grade DHEA that is micronized at prices that do appear several dollars lower than the Internet average. A phone clerk told us that Tenzing Momo started out as a tiny apothecary, one of the oldest on the West Coast, selling mainly incense and herbal products. Eventually they moved on to dietary supplements. But it's been only recently that they added DHEA supplements and a special mail-order office to handle sales for DHEA. If you want to talk with a sales clerk by phone call the toll-free ordering line at 1-800-365-9682 or the store at (206) 728-4010.

Q & A

Why is a hormone like DHEA considered a food supplement?

It is a strange situation. In the 1980s DHEA was tested as a drug treatment for obesity until the Food and Drug Administration banned its use. That's when the hormone underwent a metamorphosis of sorts and magically resurfaced as a "natural" food supplement. The law allows for the sale of dietary supplements made from natural substances over-the-counter. DHEA is a "natural" substance so it squeaks in under that heading. Some experts would like to see DHEA removed from over-the-counter formulations and sold only as a prescription drug. Others contend that the substance is safe if taken in small doses and under a doctor's supervision. It is possible the FDA may decide to change its classification.

What is the difference between DHEA in prescriptions and the DHEA bought in pharmacies or health food stores?

The first is a drug and the latter is a natural food supplement, a distinction that could prove critical if DHEA becomes a commonly used medical therapy. GL701, a pharmaceutical preparation that contains DHEA as the active ingredient, is manufactured by the biotechnology company Genelabs according to FDA regulations for purity and content or what in the pharmaceutical industry are referred to as GMPs. (All drugs must meet FDA's GMPs or Good Manufacturing Practices.) That means each and every tablet contains exactly the same dose at the same concentration. Unfortunately, no laws govern the manufacture of natural food supplements so there is no way to determine if the dose you are purchasing is the dose you are getting.

I've heard commercials for a DHEA supplement with yohimbe. What do experts think of this combination?

Yohimbe, an herbal product made from the bark of a tree native to western parts of Africa such as Cameroon, Gabon, and Congo, is purported to be an aphrodisiac. But studies have yet to prove its effectiveness and, more importantly, its side effects can be dangerous. Noted herbal expert Varro E. Tyler, a professor of pharmacognosy at Purdue University, warns that yohimbe doesn't mix well with certain foods and medications. A monoamine oxidase inhibitor, yohimbe can interact negatively with tyramine-containing foods (aged cheeses, liver, red wine), and over-the-counter diet aids or nasal decongestants that contain phenylpropanolamine. People with diabetes, low blood pressure, and heart, liver, or kidney disease would also do well to avoid this herb.

Is DHEA available in a "patch" formulation like estrogen and testosterone?

So far, only hormones such as estrogen and testosterone are available in transdermal patch formulations. (The latest testosterone patch, Androderm, is relatively new; it was approved by the FDA for use in men only in late 1995.) Since these patches offer gradual release of hormones that mimic the body's natural circadian rhythms it's not unrealistic to expect that one day a DHEA patch might become available if clinical trials confirm that DHEA replacement is a beneficial therapy for the vast majority of people.

I thought it was illegal to market medical tests to consumers without government approval. Are saliva tests government approved? If not, are the results that come by mail reliable?

Labs that do saliva hormone assays are governed by strict laboratory licensing guidelines. But some might say the saliva assay falls into a kind of legal "gray area" when it comes to government regulations. Normally, the language of the law requires that companies marketing a test or medical device to diagnose or treat a condition provide proof that the device works. (For instance, a few years ago the FDA approved an over-the-counter cholesterol test that they deemed safe and effective.) But consumers don't use saliva tests as either a treatment or a diagnostic tool. In addition, the test is noninvasive and so is safe to implement. More importantly, the two companies that market these tests are considered highly reliable by practitioners in the field. In fact, the test kit they send to consumers is the same test kit sent to physicians, clinics, and other health care facilities. The price tag is just different.

A hormone supplement I was looking at mentioned a "botanical building block" as the source for DHEA. What is that?

This is another way of listing what some companies refer to as a "natural plant source" of DHEA, usually wild yam. While it's nice to think that there might be something in nature that could be safely converted into DHEA, the body doesn't have the equipment to break down plant sterols (cholesterol) and build them into androgen hormones such as DHEA. In other words, you're wasting your money. These products do not contain DHEA and your body won't receive any benefits from what they do contain.

Since the body converts DHEA into DHEAS (DHEA sulfate), wouldn't it make better sense for companies to manufacture DHEAS supplements?

Good question. Experts say that oral administration of DHEAS is not plausible since this form of the hormone could be broken down by acids in the stomach. Conversely, DHEA is absorbed into the blood and brought to the liver where it can be converted into DHEAS.

Why don't doctors who use alternative medicine treatments do tests to confirm anecdotal evidence about DHEA?

While they have yet to test DHEA, many alternative medicine practitioners favor a method of study called "outcomes research." This type of research involves looking at patient medical records (from your patient and from other practices) and trying to figure out which therapies are beneficial and which don't seem useful. It's not as expensive or as time consuming as clinical trials. The downside to this method, however, is that there is no way to account for any other variables such as other treatments or medical problems that could be influencing the outcome to a specific therapy. Many alternative medicine specialists test treatments on themselves before giving them to a patient as a means of testing outcome.

Are most researchers convinced that DHEA holds promise as a treatment for many disorders, or are there some skeptics in the bunch?

While many hormones do have several functions,

one group of Italian researchers express skepticism that a single steroid such as DHEA can cause weight loss in obese animals, correct blood sugar in diabetic animals, boost the activity of the immune system, lower blood cholesterol levels, ward off cancerous tumors, and improve the memory of aged mice. They accept that some of the benefits of DHEA probably can be attributed to the fact that this androgen is converted to testosterone and indirectly to estrogen. They wonder if DHEA is converted to "still other metabolites" that may each be responsible for one or more of the effects mentioned above. In other words, not every one is gung ho about DHEA. Moreover, the animal studies used to test DHEA's merits all use extraordinarily large doses of DHEA. That's why trying to raise the level of DHEA in your blood naturally, which we'll discuss in the next chapter, together with healthful lifestyles habits, may be the smartest step for now.

Chapter 12

Raising DHEA Levels the Natural Way

If you're leery of supplements or simply want to wait until researchers are more certain about proposed benefits, there are ways to help boost levels of DHEA in the blood naturally. Granted these strategies may not have as strong an impact as supplements. But they aren't risky. Even better, they all employ behaviors that are beneficial to health for a number of reasons. That is, they don't just raise DHEA levels but lower risk for chronic illnesses such as heart disease and cancer and promote longevity. Indeed, good health is the result of a combination of factors that perhaps includes DHEA but is also dependent on your genetic blueprint as well as the way you take care of your body.

Learn how to manage stress.

Preliminary findings show that blood levels of DHEA are depressed in situations of stress and during illnesses, which, of course, put major stress on the body. The stress of surgery has also been shown to lower blood levels of this adrenal hormone. So if you want to keep DHEA levels as high as possible, it's important to minimize stress. How much stress is harmful? That's difficult to say. It's not easy to quantify how much and what types of stress damage health. Then again, not everyone reacts to the same stressors in the same way.

Nevertheless, what is common knowledge is that excessive amounts of stress do trigger physiological changes that if left unchecked may wreak havoc on body tissues. (Some experts say that excess stress is what you experience when responsibilities and demands exceed what you are capable of humanly accomplishing.) During acute stress, the body floods the system with the hormones epinephrine (adrenaline) and cortisol in an effort to prepare you to fight or manage extraordinary demands. Those chemical messengers instigate physiologic changes: increased heart rate, increased blood pressure, rapid breathing.

It's all well and good that this mechanism kicks into effect when you're confronted with acute danger. But this kind of response is destructive if you experience it on a routine basis. And chances are you probably do. Lifestyles today are hectic. Work demands, family responsibilities, and financial pressures seem to contribute to the problem. How bad is it? You're the only one who can gauge that answer. Hence, you're going to need to tune in to how stress impacts on your life and when situations seem overly stressful, you'll need to try and diffuse that tension. Meditation, yoga, and exercise are excellent ways to relieve stress.

Quick stress relief

Obviously you can't take time out to do yoga or take a brisk walk every time something threatens to make you lose your cool. Experts at the Mayo Clinic have a quick simple step stress-buster strategy that you can implement anytime, anywhere: while stuck in traffic, in the express checkout line when the joker in front of you has a loaded shopping cart. You get the picture. Life is full of little frustrations. Don't let them "stress you out."

1. Inhale slowly to a count of four.
   Imagine the inhaled, warm air flowing to all parts of your body.
2. Pause.
3. Exhale slowly, again counting to four: Imagine the tension flowing out.
4. Pause, then begin again. Repeat several times.

More stress: Drowning in a caffeine routine

Meditation, yoga, or exercises such as the Mayo Clinic stress-buster above aren't likely to yield dramatic relaxation benefits if you're overdoing it on a popular over-the-counter drug: caffeine. Most people forget that this powerful stimulant raises the heart rate in much the same way as stress. How much is too much? That turns out to be pretty individual. Estimates are that it takes about 200 to 240 milligrams of caffeine to provide druglike stimulatory effects. But caffeine sensitivity tends to vary from person to person.

Scientists at the Food Additive Chemistry Evaluation Branch of the FDA provide the following figures of average caffeine content of a variety of popular foods. Use these numbers to add up the amount of caffeine in your diet. If it's excessive, you might want to consider switching to decaffeinated alternatives.

Caffeine levels of Foods and Beverages

Item	Milligrams of caffeine
Coffee (5 oz. cup)	
Brewed, drip method	115
Brewed, percolator	80
Instant	65
Decaffeinated, brewed	3
Decaffeinated, instant	2
Tea (5 oz. cup)	
Brewed, major U.S. brands	40
Brewed, imported brands	60
Instant	30
Iced (12 oz. glass)	70
Cola-style soft drinks, 12 oz. can	38
Hot chocolate, 5 oz. cup	4
Chocolate milk, 8 oz. glass	5
Dark chocolate, semisweet, 1 oz.	20
Milk chocolate, 1 oz.	6

Source: Adapted from *FDA Consumer,* a publication of the Food and Drug Administration (FDA.) Reprinted with permission.

Exercising one of your best options

One former director at the National Institute on Aging was fond of stating the strong case for exercise this way: "If exercise could be packaged into a pill, it would be the single most widely prescribed, and beneficial, medicine in the nation." When it comes to DHEA levels, he might just be right. Speculation is that exercise may boost levels of this hormone in the blood. So if you want to live longer and healthier one of the best natural strategies you can employ is physical activity. And it doesn't have to be a grueling workout. New research finds distinct health benefits for people who engage in even a mildly active lifestyle such as a thirty-minute after-dinner walk each day. Experts recommend that you *accumulate* thirty minutes of *moderate* intensity activity (see the chart below) on most if not all days.

Figuring out what kind of exercise

A study of college alumni done by researchers at Stanford University in 1991 was one of the first reports to forge a strong link between physical activity and longevity. Researchers studying the health habits of over 16,000 alumni from Harvard noticed that men who routinely burned 2,000 calories per week in exercise (walking, playing sports) had a 28 percent lower death rate than their comrades who were less active. Age didn't seem to matter. Neither did blood pressure nor smoking habits. The active men were simply living longer as a group because they burned more energy in activity. In the most recent Surgeon General's report on Physical Activity and Health (1996) experts updated that advice. This time they didn't talk in terms of calories but in terms of exercise intensity. To gain health benefits from exercise you'll need to incorporate some of the following activities into your daily routine.

EXAMPLES OF MODERATE AMOUNTS OF ACTIVITY

Washing and waxing a car for 45–60 minutes
Washing windows or floors for 45–60 minutes
Playing volleyball for 45 minutes
Playing touch football for 30–35 minutes
Gardening for 30–45 minutes
Wheeling self in wheelchair for 30–40 minutes
Walking 1¾ miles in 35 minutes (20-minute mile)
Basketball (shooting baskets) for 30 minutes
Bicycling 5 miles in 30 minutes
Dancing fast (social) for 30 minutes

Pushing a stroller 1½ miles for 30 minutes
Raking leaves for 30 minutes
Walking 2 miles in 30 minutes (15-minute mile)
Water aerobics for 30 minutes
Swimming laps for 20 minutes
Wheelchair basketball for 20 minutes
Basketball (playing a game) for 15–20 minutes
Bicycling 4 miles in 15 minutes
Jumping rope for 15 minutes
Running 1½ miles in 15 minutes (10-minute mile)
Shoveling snow for 15 minutes
Stairwalking for 15 minutes

Note: Activities at the top of the list are less vigorous so need to be done for longer amounts of time. At the bottom of the list activities are more vigorous and can be pursued for less time.

Source: At-A-Glance companion document to *Physical Activity and Health: A report of the Surgeon General*, August 1996.

Pay attention to what you eat.

Studies suggest that the "hot" flashes some women experience during menopause may be triggered by alcohol and certain foods such as spicy dishes or coffee, chocolate, and other caffeine-rich items. New reports find that foods containing soy protein may boost the body's estrogen levels. Yet, no matter what the specific findings may detail, there's no doubt that what you eat can have a huge impact on health and well-being. There's no proof that healthy eating can directly raise DHEA levels. But indirectly it can have an impact on all the diseases—heart disease, cancer, diabetes, obesity—that DHEA is proposed to help. One of those indirect relations involves soy foods and estrogen.

Epidemiologists who study the incidence of disease in different populations have long known that Asian women, whose diets contain large quantities of soy foods, have lower risk for certain types of cancer. Now a soon-to-be-published study finds that adding foods that contain soy protein to a low-fat diet regimen seems to help increase bone mineral density in postmenopausal women. Researchers at the University of Illinois reported their findings at an 1996 international symposium. The protective culprit: isoflavones. Researchers explain that isoflavones behave like weak estrogens in the body and speculate that soy products may be an alternative to estrogen replacement, particularly for women at risk for breast cancer or who for other reasons are unable to tolerate estrogen supplements.

Researchers added soy protein to baked goods and beverages so that women ended up with forty grams of soy protein each day. Soy foods vary in their isoflavone content but estimates are that one cup of soy milk or one-half cup of miso, soy nuts, tempeh, soybeans, or textured vegetable protein contains thirty-five to forty milligrams of isoflavones.

Tips for men and women

Advice about soy protein may not seem like it applies to men but researchers plan to explore that issue at some point in the future. In the meantime, you can implement some general strategies to make your diet healthier whether or not you are a man or woman. The 1995 Dietary Guidelines for Americans, issued by U.S. Department of Health and Human Services, is the best place to start.

• *Eat a variety of foods* to get the energy, protein, vitamins, minerals, and fiber you need for good health.

• *Balance the food you eat with physical activity, maintain or improve your weight* to reduce your chances of having high blood pressure, heart disease, a stroke, certain cancers, and the most common kind of diabetes.

• *Choose a diet with plenty of vegetables, fruits, and grain products* that provide needed vitamins, minerals, fiber, and complex carbohydrates, and can help you lower your intake of fat.

• *Choose a diet low in fat, saturated fat, and cholesterol* to reduce your risk of heart attack and certain types of cancer. Because fat contains over twice the calories on an equal amount of carbohydrates or protein, a diet low in fat can help you maintain a healthy weight.

• *Choose a diet moderate in sugars.* A diet with lots of sugars has too many calories and too few nutrients for most people and can contribute to tooth decay.

• *Choose a diet moderate in salt and sodium* to help reduce your risk of high blood pressure.

• *If you drink alcoholic beverages, do so in moderation.* Alcoholic beverages supply calories, but little or no nutrients. Drinking alcohol is also the cause of many health problems and accidents and can lead to addiction.

Harness the "placebo" effect

Any discussion of living longer and healthier in the nineties and beyond can no longer ignore the profound impact that the mind can have on health. For example, studies show that meditation can help lower blood pressure. And at once ultraconservative medical schools such as Harvard and Columbia, physicians are employing mind/body healing techniques in the care and treatment regimens for almost every type of illness. Indeed, in a 1995 report in the journal *Mind/Body Medicine,* Harvard researcher Herbert Benson talks about the importance of a balance between mind and body therapy in health care. In a nutshell, he contends that health and well-being can be maximized if people balance traditional health care (pharmaceuticals, surgery, and procedures) with self-care. Too many of us rely far too much on pharmaceuticals and surgery and procedures and forget how profound an influence we can have on our own health just by harnessing a healthy outlook.

Indeed, in the final analysis, chances are we are all going to live longer. For example, during the 1980s the number of centenarians (100 years old or older) in the United States jumped 160 percent. By the year 2040, if the estimates of many demographers are accurate, there could be as many as forty million Americans aged eighty-five or older. By the year 2050 the centenarian population could swell to number 500,000 to four million people. If you want those twilight years to be enjoyable keep working on building the kind of lifestyle that promotes good health. It doesn't matter if that includes DHEA supplements or the natural strategies above. If you take care of your health that will pay off in improved well-being and perhaps the longest lifespan that your genetic blueprint can allow.

Putting a longevity lifestyle into action

After looking over the current drugs, pills, and treatments proposed to slow aging, experts at the NIA say the bottom line is that none are a sure thing. Still, they admit that even though no one substance can extend life, your chances of staying healthy and living long can be enhanced with the right type of longevity promoting lifestyle habits. That list includes these (ten) NIA tips for healthy aging:

• Eat a balanced diet, including five helpings of fruits and vegetables a day. (See the DHEA Food Program in Appendix I for more specific details.)

• Exercise regularly (check with your doctor before starting an exercise program.)

• Get regular health checkups.

• Don't smoke (it's never too late to quit).

• Practice safety habits at home to prevent falls and fractures. Always wear your seat belt in a car.

• Stay in contact with family and friends. Stay active through work, play, and community.

• Avoid overexposure to sun and cold.

• If you drink, moderation is the key. When you drink, let someone else drive.

• Keep a positive attitude toward life. Do things that make you happy.

Q & A

You mentioned that hot flashes can be triggered by certain foods. Is there anything else I can do to prevent these annoying bursts of internal heat? What about DHEA supplements?

While hot flashes are tied in some way to hormonal fluctuations, there are no studies that have tested DHEA as a treatment. Researchers know only that acute estrogen withdrawal, which can occur at any age, somehow upsets the body's temperature-regulating mechanisms. The precise cause of vasomotor flushes (hot flashes) remains elusive. However, experts suggest that it may be possible to control flashes by controlling environmental factors that seem to provoke them. Diet, as mentioned above, is one factor. Emotional distress and heat—hot weather or even a warm room—are two others. Good temperature-control strategies are simple. If you are careful to adjust the thermostat into the cool range in your car, in the office, and at home that might ward off trouble. Learning to "layer" your clothing is another alternative. Several thin layers can keep you warm in a too cold supermarket but can be easily peeled off as you step out into a hotter outside temperature. To relieve emotional stress or just for a "sense of well-being" set up a regular exercise regimen. Studies show that regular exercise may help minimize vasomotor symptoms. An added bonus: It may also improve the quality of your sleep, something that can decline in menopausal years.

You mention caffeine is a stimulant and so might be stressful. But isn't it true that caffeine can cause even more serious damage to health by promoting cancer and heart disease?

While much speculation and debate surrounds the issue of the safety of caffeine (sporadic reports pop up linking the stimulant to everything from heart disease to fibrocystic breast disease to pancreatic cancer), consensus among experts is that caffeine is basically safe in moderation. That is, one or two small cups of coffee per day. You'll have to decide for yourself if the stimulant qualities of this "moderate" dose are worth it in light of the added stress they could put on your body.

I've heard that "type A" behavior puts you at risk for heart disease and other illnesses. What exactly does that mean?

Experts define "type A" behavior as that characterized by aggressiveness, a need to compete, and rushing against the time clock. Some research seems to suggest a connection between this type of behavior in men and risk of coronary artery disease. But studies in women don't find a link. Maybe women with "type A" behavior are dealing with their stress more effectively. Or maybe there simply is no connection. But this doesn't rule out stress as a risk factor for heart disease or other illnesses. Too much stress is definitely unhealthy.

The DHEA Food Program

In the past, one group of researchers tried diligently to link healthy DHEA levels with a vegetarian diet. It's not such a wild concept when you consider that studies show vegetarians, as a group, tend to be leaner and often healthier than nonvegetarians. But even though the study couldn't find a connection between diet and DHEA, that doesn't mean a smart eating plan won't influence health. Indeed, the National Cancer Institute estimates that as much as 35 percent of all cancer deaths are diet-related. More than 132 different studies have linked specific foods or the antioxidant nutrients they contain to a lower risk for certain types of cancer. And for years, the American Heart Association and the National Heart, Lung, and Blood Institute have both promoted low-fat diets as a way to help lower the risk of heart disease. Consider, too, that preliminary findings suggest that certain components of the foods you eat, nutrients and phytochemicals, may enhance immune function. So why not couple your DHEA regimen (supplement or natural strategies to boost levels of the hormone) with a healthful style of eating. The best place to start is with the Department of Agriculture's Food Guide Pyramid.

FOOD GUIDE PYRAMID

The Food Guide Pyramid emphasizes foods from the five major food groups shown in the three lower sections of the Pyramid. Each of these food groups provides some, but not all, of the nutrients you need. Foods in one group can't replace those in another. No one food group is more important than another—for good health you need them all.

Some pointers

At the base of the Food Guide Pyramid are breads, cereals, rice, and pasta—all foods from grains. You need the most servings of these foods each day.

The next level includes foods that come from plants—vegetables and fruits. Most people need to eat more of these foods for the vitamins, minerals, and fiber they supply.

On the third level, two stories up from the base, are two groups of foods that come mostly from animals: milk, yogurt, and cheese; and meat, poultry, fish, dry beans, eggs, and nuts. These foods are important for protein, calcium, iron, and zinc.

The small tip of the Pyramid shows fats, oils, and sweets. These are foods such as salad dressings and oils, cream, butter, margarine, sugars, soft drinks, candies, and sweet desserts. These foods provide calories and little else nutritionally. Most people should use them sparingly.

FOOD GUIDE PYRAMID
A GUIDE TO DAILY FOOD CHOICES

> **Key**
>
> • Fat (naturally occurring   ▼ Sugars (added)
>   and added)
>
> These symbols show fats, oils, and added sugars in foods.

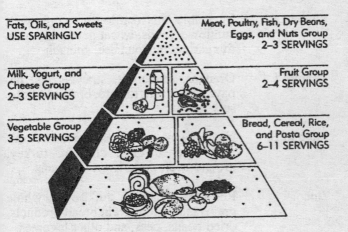

Fats, Oils, and Sweets
USE SPARINGLY

Meat, Poultry, Fish, Dry Beans,
Eggs, and Nuts Group
2–3 SERVINGS

Milk, Yogurt, and
Cheese Group
2–3 SERVINGS

Fruit Group
2–4 SERVINGS

Vegetable Group
3–5 SERVINGS

Bread, Cereal, Rice,
and Pasta Group
6–11 SERVINGS

Source: U.S. Department of Agriculture/U.S. Department of Health and Human Services

Immune-Boosting Foods

Once you've got the Pyramid approach to eating nailed down, you can branch out and think about food selections in terms of how they may enhance immune function. Preliminary findings suggest certain nutrients might boost effectiveness of some of the key players involved in defending the body against foreign invaders.

Beta-carotene Carrots, sweet potato, winter squash, turnip greens, cantaloupe, mango, papaya, apricots.

Vitamin E Vegetable oils (safflower, sunflower), sunflower seeds, wheat germ, almonds, margarine, mayonnaise, spinach

Vitamin C Oranges, grapefruit, strawberries, papaya, kiwi fruit, tomato, broccoli, brussels sprouts

Vitamin B$_6$ Navy beans, potatoes, bananas, salmon, canned tuna, chicken, turkey, ground beef, pork, liver, soybeans.

Zinc Red meat, seafood, milk, eggs, whole grains and whole-grain bread products, dried beans, peas, and other legumes.

Appendix II

Other Resources

Promising as the hundreds and hundreds of studies with DHEA appear, this hormone is not, nor will it ever be, the sole substance responsible for good health. Neither is it the only therapy or treatment scientists are studying for lupus, heart disease, weight control, AIDS, or the number of other ailments discussed in this book. If you want to keep up with the latest research developments for a specific illness, check periodically with the following organizations. Many have toll-free numbers specifically geared toward responding to consumer concerns. And since most also have access to a plethora of ongoing research projects that focus on their own specific disease category or health concern, chances are you'll be able to put new findings into proper perspective.

Appendix II: Other Resources

AIDS

CDC HIV, & AIDS Hotline
Centers for Disease Control and Prevention
U.S. Public Health Service
Atlanta, GA
1-800-342-AIDS

What they provide: Consumer information on the prevention and spread of AIDS. Does not take requests by mail.

AGING

American Geriatrics Society
770 Lexington Avenue, Suite #300
New York, NY 10021
(212) 308-1414
1-800-677-9944 (Book orders only)
Internet address: http://americangeriatrics.org

What they provide: Geared to professionals. But consumers can write away for position papers on controversial topics such as physician-assisted suicide; there is no position paper on DHEA. Callers to the toll-free number can order *"The American Geriatrics Society's Complete Guide to Aging and Health* at a 35 percent discount. Shipping costs are not included. *Note:* Internet site is "hyperlinked" to other sites on aging.

National Institute on Aging
Information Center
P.O. Box 8057
Gaithersburg, MD 20898
(301) 496-1752
(301) 589-3014 (Fax)
1-800-222-2225
Internet address: http://nih.gov/nia
E-mail: niainfo@access.digex.net

What they provide: Brochures and booklets about different facets of aging: menopause, sexuality in later life, health quackery, getting a good night's sleep.

ALZHEIMER'S DISEASE

Alzheimer's Disease Association
919 North Michigan Avenue, Suite #1000
Chicago, IL 60611
(312) 335–8700
1–800–272–3900
Internet address: http://www.alz.org
E-mail address: info@alz.org

What they provide: Brochures, books, and videos on the disease. Special packets for: general interest, newly diagnosed patients, caregivers. Also provides referrals to local chapters.

Alzheimer's Disease Education and Referral Center
P.O. Box 8250-JML
Silver Spring, MD 20907
(301) 495-3311
1-800-438-4380
Internet address: http://www.alzheimers.org/adear
E-mail address: adear@alzheimers.org

What they provide: Information for consumers and professionals. Fact sheet and yearly updates on progress in research. Referrals to government-sponsored research centers. (Service of the National Institute on Aging.)

ARTHRITIS

Arthritis Foundation
1314 Spring Street
Atlanta, GA 30309
1-800-283-7800

What they provide: Interactive recorded messages about different types of arthritis. One free brochure per phone call. Referrals to local chapter; subscription information for *Arthritis Today* magazine.

National Arthritis and Musculoskeletal and Skin Diseases
Information Clearinghouse
Box AMS
9000 Rockville Pike
Bethesda, MD 20892
(301) 495-4484

What they provide: Information on many of the more than 100 disorders that fit under the umbrella term arthritis and information on skin, muscle, and bone disorders. Brochures can be ordered via phone or you can choose to talk with an information specialist.

CANCER

American Cancer Society
1599 Clifton Road N.E.
Atlanta, GA 30329
(404) 320-3333
1-800-ACS-2345

What they provide: Referral to local chapters. Brochures and information about cancer and strategies that can help prevent various types of cancer.

American Institute for Cancer Research
1759 R Street, N.W.
Washington, D.C. 20009
(202) 328-7744
1-800-843-8114 Nutrition Hotline
Internet address: http://www.aicr.org

What they provide: Newsletter and numerous information brochures about various types of cancer as well as diet and lifestyle strategies to ward off cancer.

National Cancer Institute
at the National Institutes of Health
9000 Rockville Pike
Building 31, Room 10A24
Bethesda, MD 20892
(301) 496-5583
1-800-4-CANCER
Internet address: http://www.icicc.nci.nih.gov

What they provide: "What you need to know" series on a variety of types of cancer. Brochures and booklets on cancer prevention. Up to twenty different brochure requests are free; more than that amount requires a prepaid shipping and handling charge. Information for health professionals also.

DIABETES

American Diabetes Association
1660 Duke Street
Alexandria, VA 22314
(703) 549-1500
1-800-ADA-3472 national center
1-800-342-2383 for local affiliate

What they provide: Brochures and literature about diabetes. Local chapters offer physician referrals (endocrinologists) who treat diabetes. Members receive a complimentary subscription magazine. (ADA has no official stand on DHEA. They don't take stands on any treatments until clinical trials with people show they are valid.)

GENERAL HEALTH INFORMATION

National Health Information Center
P.O. Box 1133
Washington, D.C. 20013-1133
(202) 429-9091 in D.C.
1-800-336-4797
Internet address: http://nhic-nt.health.org
E-mail: nhicinfo@health.org

What they provide: Referrals to national organizations or government sources for information on a wide variety of diseases. Their computer database has listings for over 12,000 nonprofit health groups and federal centers. (File this number away so that if future research links DHEA to other illnesses you'll know how to find more information.)

Food and Drug Administration (FDA)
Office of Consumer Affairs, HFE-88
5600 Fishers Lane
Rockville, MD 20857
(301) 827-4420
Internet address: http://www.fda.gov

What they provide: Health information for consumers and professionals. This service routes your call to the proper FDA office. Also takes calls on adverse reactions to new drugs and answers questions about medical devices.

HEART DISEASE

American Heart Association
7320 Greenville Avenue
Dallas, TX 75231
(214) 373-6700

What they provide: Referrals to local chapters. Brochures and educational materials on atherosclerosis and heart disease. Information on AHA Step 1 and Step 2 low-fat diets. (AHA has no official stand on DHEA.)

National Heart, Lung, and Blood Institute
at the National Institutes of Health
Information Center
P.O. Box 30105
Bethesda, MD 20824
(301) 951-3260

What they provide: Brochures and information about atherosclerosis, heart disease, and low-fat diets. Special booklet for women on heart disease.

LUPUS

Lupus Foundation of America
1300 Piccard Drive, Suite #200
Rockville, MD 20850
(301) 670-9292
1-800-558-0121
Internet address: http://www.lupus.org/lupus

What they provide: A general-information packet about lupus and its treatment. Information about enrolling in clinical trials testing pharmacologically pure DHEA on mild to moderate cases of lupus. (See Chapter 8 for LFA's position on DHEA.)

MENOPAUSE

North American Menopause Society (NAMS)
P.O. Box 94527
Cleveland, OH 44101

Appendix II: Other Resources

1-900-370-NAMS ($1.95 minute)
Internet address: http://www.menopause.org/
E-mail: nams@apk.net

What they provide: Brochures on menopause, female
cancers, sexuality (how it can be affected at midlife
and beyond), abnormal uterine bleeding, etc. Information offered via E-mail or fax.

Appendix III

Can Hormones Reverse Aging?

It is beyond the scope of this one book to convey each and every scientific opinion about DHEA and other hormones that are being touted as antiaging remedies. But who better to give you an overview on how hormones affect aging than the aging experts—the scientists at the National Institutes of Health. Sensing the current hype about melatonin and the growing fervor surrounding DHEA these scientists decided to "go on the record" with their beliefs about hormones and life extension in a recently published NIA report. Here is that viewpoint, published verbatim (with permission).

NATIONAL INSTITUTE ON AGING

October 1996

In the 1890s, snake oil salesmen hawked elixirs to cure everything from the vapors to rheumatism. Some pitches claimed that the elixirs could regrow hair, cure intestinal woes, and make you look years younger.

Today, many people still cling to the hope that pills or injections can cure all of their ills and reverse aging. A variety of hormones are currently being mentioned as having remarkable curative and antiaging benefits. The question remains whether using these commonly touted antiaging hormone replacement therapies really will help you to stay young. To answer this question, scientists supported by and working at the National Institute on Aging (NIA) have been investigating the risks and benefits of several hormones. The hormones that receive most of the current media attention are estrogen, dehydroepiandrosterone (DHEA), human Growth Hormone (hGH), and melatonin.

What are hormones?

Hormones are chemical substances secreted by various endocrine glands into the bloodstream for delivery to a target organ where their effect is felt. These glands make up the body's endocrine system, a system that regulates growth, metabolism, and reproduction, as well as the function and well-being of many body components.

DHEA, hGH, melatonin, and estrogen levels may decline, on average, as people age. However, there is a great variability in these hormone levels, with some older individuals having very low levels and others having levels found in younger adults. This decline may not be important if these substances are needed most during growth and development. In fact, these hormones may not be needed in large amounts as a person grows older.

For example, the male and female sex hormones, testosterone and estrogen, are two of the best known examples of hormones whose levels change as we age. In women, levels of the most active form of estrogen drop off after menopause, a time when women no longer have reproductive capabilities. Estrogen replacement therapy (ERT), available since the 1940s, is associated with stopping or slowing some of the effects of estrogen loss, such as osteoporosis and heart disease, but ERT may increase the risk of uterine and breast cancer. Testosterone levels also fall off as a man ages, but these levels decline gradually and the impact on the body is not as dramatic as the loss of estrogen in women at menopause. NIA currently supports testosterone replacement studies in older men who have low testosterone levels to understand the health benefits, or risks, that might derive from testosterone supplementation. NIA also supports continuing studies of estrogen replacement therapy in older women to more completely understand its effects.

Estrogen

Estrogen is currently used by millions of women to reduce many of the health problems associated with menopause. The history of estrogen research is an informative case study of the complexity of research into hormones. After fifty years of intense study into estrogen, we still do not fully understand how it works or the implications of replacement therapy.

In the 1940s, when estrogen was first offered to menopausal women, it was given alone and in high doses. Today, after much experimentation, we know that these doses led to increased risk for uterine cancer. We have learned that estrogen replacement, under the right conditions, can be successful in combating osteoporosis and substantially reducing the risk of developing heart disease. There is also some evidence that estrogen may have beneficial effects on brain function and memory.

Given how long it has taken to make these discoveries about estrogen, it is unlikely that there will be quick and easy answers on how other hormones work. For each hormone currently being considered for its health or aging effects, the same intense scrutiny is essential to avoid unintended outcomes.

Doctors now prescribe hormone replacement therapy (HRT) to combat symptoms associated with falling estrogen levels. HRT involves administration of the female hormones estrogen and progesterone. Progesterone helps reduce the risk of uterine cancer associated with estrogen supplementation alone. The cardiovascular effects of progesterone are still not known.

Hormone treatment for menopause remains controversial. Its long-term safety and efficacy remain matters of great concern, especially in regard to the increased risk for breast cancer that is thought to be associated with HRT use. There is not enough data for physicians to suggest that HRT is the right choice for *all* women. Several large studies are currently attempting to resolve these questions, though it may take several more years before definitive answers are available. Five- to ten-year studies are typically needed to properly investigate the long-term benefits and risk associated with many substances, especially hormones.

The lesson to be learned from the ongoing struggle to define optimal uses and risks involved with estrogen supplementation is that the public should be wary of simple answers while scientists work to more completely understand the intricacies, risks, and benefits of other hormones.

DHEA

By the 1960s, scientists had observed that DHEA (dehydroepiandrosterone), a weak male hormone produced by the adrenal glands, is converted in the body to estrogen and testosterone. Most people produce peak amounts of DHEA around age thirty, at which time DHEA production begins to decline gradually. Human clinical trials, in which DHEA is given in relatively low doses to older people to increase their blood levels to those seen in young adults, have been too short in duration and have had too few participants to give definite results.

Optimally, long-term, double-blind, controlled clinical trials in people are the best way to judge if the benefits seen in small trials hold true for most people. Researchers are planning such trials but the results probably won't be available in this decade. This long-term research is necessary to determine if there are benefits to this treatment and if there are serious side effects that might show up later. Already there are indications that taking high levels of DHEA can lead to serious side effects, such as liver damage, in the short term. DHEA is a precursor for estrogen and testosterone, and excess amounts of these hormones from DHEA may be linked to increased risk for breast and prostate cancers. High testosterone levels in women can lead to excessive facial hair growth and changes in blood fats that increase the risk for heart disease.

Little is actually known about DHEA's role in the body. One unknown is the body's ability to convert DHEA into different forms, each of which may have varying effects. A second concern is that the bulk of DHEA in the body is in the form of DHEA-sulfate, which may function differently from the unsulfated form of the hormone. Other unanswered questions include whether DHEA and its sulfate simply serve as storage pools for the manufacture of estrogen and testosterone, or whether these substances also have biological properties of their own. Because of these uncertainties, scientists don't know whether DHEA is directly responsible for many of the beneficial effects claimed for it, or if the beneficial effects are the results of DHEA's conversion to estrogen and testosterone.

In 1994, Congress passed the Dietary Supplement Health and Education Act, which allows DHEA and melatonin to be sold over-the-counter as nutritional supplements as long as labels don't contain unsubstantiated health claims. However, use of DHEA is not advised until further research results determine if there are positive health effects and, if so, what is the proper dosage and form of administration of DHEA that will maximize benefits and minimize side effects.

Currently, the NIA supports studies of DHEA in animals to determine if it has "antiaging" effects. In one of these studies in animals, scientists are administering DHEA-sulfate to mice to verify earlier reports that it can block the decline of the immune system that is normally seen in older mice. Eventually, this immune system study could lead to insights into DHEA's effect on aging. Current human trials of hGH should give us more definitive answers about how this particular hormone works in people.

Human growth hormone

Human Growth Hormone (hGH) is secreted by the anterior pituitary gland and exerts an effect on protein, carbohydrate, and lipid metabolism. It plays a role in determining the rate of development of skeletal bones and many large organs in the human body. To complicate an already complex process, secretion of hGH is controlled by two substances—somatostatin, which inhibits hGH release, and growth-releasing hormone, which stimulates hGH release. The decrease in hGH with age is believed due in part to decreased levels of releasing hormone. Supplementation of the releasing hormone is of interest to scientists looking for ways to increase hGH levels.

HGH is one of several trophic factors, substances that promote the growth or maintenance of tissue (estrogen and testosterone are two others), that are being investigated in five-year clinical studies. The trophic factors' studies receive support from the NIA at nine centers throughout the United States. Five of these centers investigate the benefits and risk associated with hGH supplementation. Researchers at these centers are studying hGH to determine whether it can help strengthen muscle and bone and help reduce frailty in older people. Because NIA scientists believe that no one approach will be a "cure-all" for the effects of aging, they also are looking at hGH supplementation in conjunction with strength-training exercise and aerobic training. Results of these studies will not be available until late 1997 at the earliest.

Testing hGH in a clinical trial, on many older people over a reasonable period of time, will provide solid information on the clinical utility of hGH as well as on the risks associated with supplementation. When these trials are complete, hGH supplementation will be better understood. Currently, hGH is available by prescription for children with growth hormone deficiency, but it is not approved for use in adults.

An unregulated medical clinic setup in the early 1990s in Mexico to promote the use of hGH closed after the public learned about hGH's negative side effects and the clinic's poor monitoring of hGH injections. Some of the complications of excess hGH are significant and include development of carpal tunnel syndrome, hypertension, and diabetes. In the next few years NIA scientists should have a much better idea about the risks and benefits of hGH, so it would be wise to wait until all of the study results are in before seeking hGH shots.

Melatonin

Even less is known about the possible antiaging effects of melatonin than of DHEA or hGH. Melatonin is produced by the pineal gland, located deep inside the brain. At night, in the absence of light on the retina, the brain signals the pineal gland to release melatonin, which is thought to help induce sleep. Melatonin, present in many animals, can act on other organ systems and influence activities such as breeding behaviors, migration patterns, shedding of coats, and many other changes in various species. Low-dose melatonin supplements are being investigated in people as a means to promote sleep. Some people find it effective in the short term, but others do not, and may, in fact, find their sleep more impaired. Claims that melatonin also can affect the body's aging clock and fix the ill effects of aging are based only on severely limited and questionable laboratory studies in animals. The claims for human benefit are premature.

Synthetic melatonin, like DHEA, currently is sold over-the-counter, but the long-term and high-dosage effects of this hormone are unknown and studies of its effects on aging have not yet been conducted. The amount of melatonin in the body may decline with age in many people, but until the consequences of this decline and the effect of long-term administration of melatonin have been rigorously analyzed, its use as a long-term supplement cannot be recommended. If a person insists on using melatonin, this should be done with the knowledge and monitoring of his or her physician.

One of the best cautions about purported antiaging compounds comes from a recent critical scientific review entitled *Melatonin Madness:*

> Rene Descartes, the French philosopher and mathematician, proclaimed in the seventeenth century that the pineal gland was the "seat of the soul." Even Descartes would be astonished at the properties currently being attributed to melatonin. Continued progress in understanding the cellular and molecular actions of melatonin will hopefully remove the mystery surrounding this hormone. The cure for melatonin madness is to ignore the hyperbole and histrionics and focus instead on hypothesis testing and sound science.

Antioxidants and possible antioxidant properties of hormones

Some people say that melatonin acts as an antioxidant—a claim that is made for many alleged antiaging substances. Antioxidants are found in common foods, especially those rich in vitamins A, C, and E. They act as scavengers for free radicals, which result from normal metabolism in the body. Left "unscavenged," free radicals can cause long-term harm and degradation to the body and its cellular components. The antioxidant properties of melatonin within the body are not clearly understood and need further investigation.

Along with melatonin, vitamins that counteract the detrimental effects of free radicals also are touted as remedies for aging. While there is no question that vitamins such as A, C, and E, when consumed in foods such as carrots, oranges, and broccoli, are essential for health, taking more than the recommended daily allowance of vitamin supplements has not been shown to be beneficial. The question, "do antioxidant vitamins protect us from cancer or aging?" remains unanswered. In a 1994 study of male smokers in Finland, taking vitamin E supplements did not reduce the rate of lung cancer at all. In fact, these researchers found an 18 percent higher rate of lung cancer in men who took beta-carotene daily for five to eight years. Two recent National Institutes of Health-supported clinical trials on beta-carotene showed no benefit, and even some possible harm, due to beta-carotene supplementation. For people who are ill or who have vitamin deficiencies, supplements may be useful, but antiaging effects of high-dose vitamins have not been demonstrated. And with any substance that has claims that are too good to be true, caution is probably the best watchword.*

*Reppert S, Weaver D. (1995) Cell 83, Melatonin Madness, 1059–62.

National Institutes of Health
National Institute on Aging
October 1996
(Reprinted with permission.)

Glossary

ACTH—The shorthand term for adrenocorticothrophic hormone, a chemical produced by the pituitary gland (the "master" gland that regulates the activities of all the other endocrine glands). ACTH signals the adrenal gland to produce DHEA, testosterone, and other steroid hormones.

AIDS—The acronym stands for acquired immune deficiency syndrome, an illness that devastates the immune system leaving the body open to "opportunistic infections," infections that typically don't happen when immune defenses are intact. HIV (human immunodeficiency virus) is the name of the viral infection that causes AIDS. However, not everyone infected with HIV develops AIDS.

Epinephrine—The sympathetic nervous system alerts the adrenal gland to produce this hormone, which is also called adrenaline in response to fear, stress, or exercise. Synthetic epinephrine has long been used as a drug treatment for severe allergic reactions (anaphylactic shock), cardiac arrest, and acute asthma attacks.

Adrenals—Small, oddly shaped glands that sit atop the kidney. The adrenal medulla portion of the adrenal gland secretes the hormones adrenaline (epinephrine) and norepinephrine to help the body respond to stress; the adrenal cortex secretes cortisol and aldosterone.

Alzheimer's disease—The most common form of senile dementia; a progressive disease of the brain marked by loss of mental abilities: thinking, problem solving, and memory.

Androderm—Brand name of a transdermal patch used to deliver testosterone through the skin. It was approved for use (in men only) by the Food and Drug Administration in September 1995.

Androgen—The name given to a class of hormones (testosterone is the predominant one) that have masculinizing effects. That is, they stimulate the growth of facial and body hair, deepen the voice, increase muscle mass, etc. Androgens are produced both in the testes (in males), in the adrenal glands, and in small quantities in the ovaries (in females) prior to menopause. Androgens produced by the testes are more potent stimulators of male secondary sexual characteristics than those produced in the adrenals.

Aphrodisiac—Foods, chemicals, and substances believed to stimulate sexual desire and enhance sexual performance. While the list of "love potions" is long, there is no firm scientific evidence that any one ingredient can heighten sexual desire. On the other hand, since the brain may be the most powerful sexual organ, if a person believes strongly enough that a substance will work it just might.

Appetite suppressants—A class of drugs that cuts appetite and the desire to eat. Over-the-counter products typically contain phenylpropanolamine. Popular prescription products include fenfluramine, phentermine, and the new drug Redux. Belief is that these medications dampen appetite via influence on the hypothalamus, the part of the brain that exerts control over the sympathetic nervous system.

Atherosclerosis—A disease in which the inner walls of arteries narrow due to the buildup of plaque (cholesterol and other substances). Progressive thickening of artery walls can eventually obstruct blood flow to the heart, brain, kidneys, intestines, and the lower extremities.

Circadian rhythm—A biological pattern that is cyclical in nature for the course of a day or roughly twenty-four hours. Many physiological functions vary in a rhythmic way (menstrual cycle is approximately twenty-eight days) but most are either circadian (twenty-four hours) or twice-daily cycles. Hormones help to regulate an internal body clock that regulates these rhythms.

Coronary heart disease—Damage to the heart caused by narrowing or blockage of the arteries that supply the heart. Genetics and the disease diabetes can increase your risk of developing CHD. But other risk factors—smoking, lack of exercise, obesity, elevated blood cholesterol levels—are under your control.

Cytokine—Hormonelike protein secreted by various immune cells including the T-cells and macrophages.

Scientists have harnessed some cytokines such as inter-leukin-2 and the interferons and are testing or using them as drug therapy in the treatment of certain diseases.

Double blind study—This is a type of clinical trial in which neither the participants nor the researcher know who is receiving the treatment being tested versus who is taking a placebo. Since both researcher and participant are "blind" to the treatment this lessens the chance that results will be biased. Randomized, double-blind placebo-control studies are considered the "gold standard" of medical research.

Endocrine gland—A group of specialized cells that secrete chemical substances called hormones directly into the bloodstream for transport throughout the body. For example, the pancreas secretes the hormone insulin, which controls blood sugar. Other examples are the thyroid gland, the ovaries, and, of course, the adrenal glands, which produce DHEA and a number of other key hormones.

Endocrinologist—A physician who treats disorders involving any of the endocrine glands. That list includes more common ailments such as diabetes and thyroid disorders.

Free radicals—Naturally occurring, highly reactive substances that damage DNA and destroy cells. Researchers say that we all have an inherent ability (our genes determine this) to fight against this kind of destruction. Some nutrients (vitamins E and C) and perhaps DHEA may also have the capacity to neutralize these destructive compounds.

FSH—Biochemist shorthand for follicle stimulating hormone, a chemical secreted by the pituitary gland. FSH stimulates the maturation of follicles, the sacs in which eggs develop, in the ovaries.

Glucocorticoids—Hormones produced by the adrenal glands that affect how the body metabolizes carbohydrates. The main glucocorticoid is cortisol (hydrocortisone). When the body is subjected to major stressors (surgery, illness) the adrenal glands step up production of cortisol.

High blood pressure—Also referred to as hypertension, a silent illness marked by increased pressure of the blood in the main arteries while the body is at rest. It's usually defined as a resting blood pressure that is greater than 140 mm Hg/90mm Hg. Mild cases sometimes respond to weight-reduction and stress-reduction strategies.

High density lipoprotein—Called "HDL" for short, this floating package of lipids and protein is often referred to as the "good" cholesterol because it helps carry cholesterol out of the bloodstream to the liver, thereby leaving less opportunity for cholesterol to stick to artery walls.

Hormones—Chemical messengers that travel throughout the body regulating and coordinating various metabolic functions. Hormones can be divided into three basic categories based on their chemical structure: steroids (testosterone, DHEA), substances derived from amino acids, the building blocks of protein (thyroid hormones), and peptides. Some hormones work at the cell's surface while others (steroids) can actually enter the cell.

Hormone replacement therapy—Traditionally this term refers to the practice of replacing some of the estrogen (ERT) and progesterone lost by women in the post-menopausal years with natural or synthetic hormones. With testosterone replacement becoming more common in men (and women) and clinical trials testing the feasibility of restoring DHEA to youthful levels this term is likely to take on a broader meaning in the years ahead.

Hypothalamus—The part of the brain that regulates many basic body functions including temperature, sleep, sexual behavior, and mood or emotions. About the size of a cherry, this section of the forebrain sits directly above the pituitary gland, the "master" controller of the endocrine system.

Interferons—Protein substances secreted by a variety of body cells in order to help the body defend against viral invaders. There are three types of interferons: alpha, beta, and gamma. Alpha interferon has proven to be helpful in treating certain types of cancer including melanoma (skin) and colon cancer.

Interleukins—Chemicals secreted by the immune system's T-cells. Their role is to help recognize and mount an attack against foreign invaders. Interleukin-2 alerts the immune system that a foreign substance has breached the defenses and signals other interleukins and interferons to activate the production of T-cells and B-cells in an effort to marshal the forces necessary to mount an attack against that invader. (Interleukins belong to a larger class of immune defenders called cytokines.)

Intracrinology—A new offshoot of endocrinology concerned with studying hormone activity within the cell.

Intracrinologists are currently looking into the activity of steroid hormone DHEA in hopes they will uncover the mechanism by which it may be converted into active androgens (testosterone) and estrogens within various tissues. (Remember, belief is that much of DHEA is converted into androgens in the liver rather than target tissues.)

Libido—Sex drive, sexual urge, or instinct.

Liver—The body's "chemical factory," this large organ (at about three pounds it's the largest body organ) is responsible for regulating the levels of many internal body chemicals. With the help of the kidneys it rids the body of toxins and drugs that might otherwise accumulate in the bloodstream. Preliminary reports have raised concern that druglike doses of DHEA might overtax the liver and could lead to liver cancer.

Low-density lipoprotein—Often referred to in shorthand as "LDL," or the "bad" cholesterol. This floating package of fat (lipids) and protein contains cholesterol that can adhere along with other substances to artery walls and restrict blood flow. Speculation is that compounds in the body called "free radicals" may alter LDL particles and make them more likely to stick to artery walls.

Lupus—Short for lupus erythematosus, this disease is an autoimmune disorder in which the body's own internal defenses attack the connective tissues throughout the body; painful swelling and inflammation can be a problem. While the cause is unknown, there appears to be a strong genetic component. And speculation is that hormonal factors may also play a role since women are ten times more likely to be affected than men.

Norepinephrine—A hormone made in the adrenal gland that helps to stabilize blood pressure in the wake of acute or chronic stress. It works in tandem with adrenaline and one other hormone to aid in what most of us know as the "fight or flight" response. Norepinephrine is also synthesized in the brain.

Osteoporosis—A loss of density and gradual weakening of bone tissue that results in bones that are brittle and easier to fracture. Postmenopausal women are particularly susceptible to the disorder since they no longer produce estrogen, a hormone that helps maintain bone mass.

Pituitary—The "master" gland that regulates the activities of all the other endocrine glands.

Placebo—A harmless, inactive pill or tablet. In trials to test a new drug placebos are made to look and taste like the drug they mimic. If participants in the study can't tell which treatment they are receiving they can help eliminate bias and give researchers a better idea of the drug's effectiveness.

Placebo effect—If an individual is convinced a treatment or medication will work he or she may benefit from the therapy solely based on the impact of positive belief.

Postmenopausal—Refers to the time period following menopause (the time in a women's life when menstrual periods have stopped for at least one year). The average age for menopause in the United States is fifty-one years of age.

Steroid—A chemical compound with a characteristic structure that includes a ring of carbon atoms. All human steroids are made from the parent compound cholesterol and are manufactured in endocrine glands such as the ovary (estrogens) and the testis (androgens) and the adrenal gland (DHEA, cortisol, aldosterone). Although nearly fifty different steroids have been isolated in the adrenal gland, only a handful of them are biologically active hormones.

Testosterone—Hormone responsible for male secondary sex characteristics: hair on face and chest, deepening of voice, enlargement of penis. Testosterone plays a role in stimulating libido in men and perhaps women. It's produced by the testes in men and in small amounts by the ovaries in women. The body is also capable of converting DHEA into testosterone.

Virus—The smallest known infectious agent capable of breaching the body's defenses, viruses come in all shapes and sizes. But unlike bacteria, viruses cannot multiply on their own; like parasites they use their host (human or animal) to survive. The common cold or warts are both caused by viral infections that use their host to grow and multiply. Viruses are also responsible for more serious illnesses such as AIDS and rabies.

References

CHAPTER ONE

Davis, S. R., Burger, H. G. "Clinical review 82: Androgens and the postmenopausal woman." *Journal of Clinical Endocrinology and Metabolism* 1996; 81:2759–63.

Nestler, J. E. "DHEA: A coming of age." *Annals of the New York Academy of Sciences* 1995; 774:ix–xi.

Hornsby, P. J. "Current challenges for DHEA research." *Annals of the New York Academy of Sciences* 1995; 774:xiii–xiv.

Regelson, W., Loria, R., Kalimi, M. "Dehydroepiandrosterone (DHEA)—the 'Mother Steroid.' " *Annals New York Academy of Sciences* 1994; 719:553–63.

Birkenhager-Gillesse, E. G., Derksen, J., Lagaay, A. M. "Dehydroepiandrosterone sulphate (DHEAS) in the oldest old, aged 85 and over." *Annals New York Academy of Sciences* 1994; 719:543–52.

Regelson, W., Kalimi, M. "Dehydroepiandrosterone (DHEA)—the multifunctional steroid: Effects on the CNS, cell proliferation, metabolic and vascular, clinical and other effects. Mechanism of action?" *Annals New York Academy of Sciences* 1994; 719:564–75.

National Cancer Institute, Chemoprevention Branch and Agent Development Committee. "Clinical development plan: DHEA analog 8354." *Journal of Cellular Biochemistry,* Supplement 1994; 20:141–46.

Kelloff, G. J., Crowell, J. A., et al. "Strategy and planning for chemopreventive drug development: Clinical Development Plans." *Journal of Cellular Biochemistry,* Supplement 1994; 20:55–299.

Drucker, W. D., Blumber, J. M., et al. "Biologic activity of dehydroepiandrosterone sulfate in man." *Journal of Clinical Endocrinology and Metabolism* 1972; 35:48–54.

Martin, D. W., Mayes, P. A., Rodwell, V. W. and Granner, D. K. *Harper's Review of Biochemistry,* 20th edition. Lange Medical Publications, 1985.

CHAPTER TWO

Ravaglia, G., Forti, P., et al. "The relationship of dehydroepiandrosterone sulfate (DHEAS) to endocrine-metabolic parameters and functional status in the oldest-old. Results from an Italian study on Healthy free-living over-ninety-year-olds." *Journal of Clinical Endocrinology and Metabolism* 1996; 81:1173–78.

Manton, K. G., Stallard, E. "Longevity in the United States: Age and sex-specific evidence on life span limits from mortality patterns 1960–1990." *Journal of Gerontology* 1996; 51A:B362–75.

Finch, C. E., Pike, M. C. "Maximum life span predictions from the Gompertz mortality model." *Journal of Gerontology* 1996; 51A:B183–94.

Touitou, Y. "Effects of ageing on endocrine and neuroendocrine rhythms in humans." *Hormone Research* 1995; 43:12–19.

Birkenhager-Gillesse, E. G., Derksen, J., Lagaay, A. M. "Dehydroepiandrosterone sulphate (DHEAS) in the oldest old, aged 85 and over." *Annals New York Academy of Sciences* 1994; 719:543–52.

Morales, A. J., Nolan, J. J., Nelson, J. C., and Yen, S. C. "Effects of replacement dose of dehydroepiandrosterone in men and women of advancing age." *Journal of Clinical Endocrinology and Metabolism* 1994; 78:1360–67.

Belanger, A., Candas, B., et al. "Changes in serum concentrations of conjugated and unconjugated steroids in 40- to 80-year-old men." *Journal of Clinical Endocrinology and Metabolism* 1994; 79:1086–90.

Regelson, W., Kalimi, M. "Dehydroepiandrosterone (DHEA)—the multifunctional steroid: Effects on the CNS, cell proliferation, metabolic and vascular, clinical and other effects. Mechanism of action?" *Annals New York Academy of Sciences* 1994; 719:564–75.

CHAPTER THREE

Barrett-Connor, E., Ferrara, A. "Dehydroepiandrosterone, Dehydroepiandrosterone sulfate, obesity, waist-hip ratio, and noninsulin-dependent diabetes in postmenopausal women: The Rancho Bernardo study." *Journal of Clinical Endocrinology and Metabolism* 1996; 81:59–64.

Moghetti, P., Tosi, F., et al. "The insulin resistance in women with hyperandrogenism is partially reversed by antiandrogen treatment: evidence that androgens impair insulin action in women." *Journal of Clinical Endocrinology and Metabolism* 1996; 81:952–60.

Marin, P., Lonn, L., et al. "Assimilation of triglycerides in subcutaneous and intraabdominal adipose tissues in vivo in men: effects of testosterone." *Journal of Clinical Endocrinology and Metabolism* 1996; 81:1018–22.

Nelson, L. H., Tucker, L. A., et al. "Diet composition related to body fat in a multivariate study of 203 men." *Journal of the American Dietetic Association* 1996; 96:771–7.

Morales, A. J., Nolan, J. J., Nelson, J. C., and Yen, S. C. "Effects of replacement dose of dehydroepiandrosterone in men and women of advancing age." *Journal of Clinical Endocrinology and Metabolism* 1994; 78:1360–67.

Jakubowicz, D. J., Nusen, A. B., et al. "Disparate effects of weight reduction by diet on serum dehydroepiandrosterone-sulfate levels in obese men and women." *Journal of Clinical Endocrinology and Metabolism* 1995; 80:3373–76.

Nakashima, N., Haji, M., et al. "Effect of dehydroepiandrosterone on glucose uptake in cultured human fibroblasts." *Metabolism* 1995; 44:543–48.

Tchernof, A., Despres, J. P., et al. "Reduced testosterone and adrenal C19 steroid levels in obese men." *Metabolism* 1995; 44:513–19.

Svec, F., Hilton, C. W., et al. "The effect of DHEA given chronically to Zucker rats." Proceedings of the Society of Experimental Biology and Medicine 1995; 209:92–97.

Haffner, S. M., Valdez, R. A., et al. "Decreased testosterone and dehydroepiandrosterone sulfate concentrations are associated with increased insulin and glucose concentrations in nondiabetic men." *Metabolism* 1994; 43:599–603.

Berdanier, C. D., Parente, J. A., et al. "Is dehydroepiandrosterone an antiobesity agent?" *The FASEB Journal* 1993; 7:414–419.

Wright, B. E., Browne, E. S., et al. "Divergent effect of

dehydroepiandrosterone on energy intakes of Zucker rats." *Physiology & Behavior* 1993; 53:39–43.

Williams, D. P., Boyden, T. W., et al. "Relationship of body fat percentage and fat distribution with dehydroepiandrosterone sulfate in premenopausal females." *Journal of Clinical Endocrinology and Metabolism* 1993; 77:80–85.

Buffington, C. K., Pourmotabbed, G., et al. "Case report: Amelioration of insulin resistance in diabetes with dehydroepiandrosterone." *American Journal of the Medical Sciences* 1993; 306:320–24.

Usiskin, K. S., Butterworth, S., et al. "Lack of effect of dehydroepiandrosterone in obese men." *International Journal of Obesity* 1990; 14:457–63.

Welle, S., Jozefowicz, R., et al. "Failure of dehydroepiandrosterone to influence energy and protein metabolism in humans." *Journal of Endocrinology and Metabolism* 1990; 71:1259–1264.

Nestler, J. E., Barlascini, C. O., et al. "Dehydroepiandrosterone reduces serum low density lipoprotein levels and body fat but does not alter insulin sensitivity in normal men." *Journal of Clinical Endocrinology and Metabolism* 1988; 66:57–61.

CHAPTER FOUR

Marin, P., Lonn, L., et al. "Assimilation of triglycerides in subcutaneous and intraabdominal adipose tissues in vivo in men: effects of testosterone." *Journal of Clinical Endocrinology and Metabolism* 1996; 81:1018–22.

Dandona, P., Thusu, K., et al. "Oxidative damage to DNA in diabetes mellitus." *Lancet* 1996; 347:444–5.

Ettinger, B., Friedman, G. D., et al. "Reduced mortality associated with long-term postmenopausal estrogen therapy." *Obstetrics & Gynecology* 1996; 87:6–12.

References

O'Keefe, J. H., Nelson, J., et al. "Lifestyle change for coronary artery disease." *Postgraduate Medicine* 1996; 99:89–106.

U.S. Department of Health and Human Services. "Physical activity and health: a report of the Surgeon General." Atlanta: U.S. Department of Health and Human Services, Public Health Service, Centers for Disease Control and Prevention, 1996.

NIH Consensus Development Panel on Physical Activity and Cardiovascular Health. "Physical activity and cardiovascular health." *Journal of the American Medical Association* 1996; 276:241–6.

Editors, Mayo Clinic Health Letter. "Heart Matters: 8 ways to lower your risk of a heart attack or stroke." Mayo Foundation for Medical Education and Research, 1996.

Haffner, S. M., Newcomb, P. A., et al. "Relation of sex hormones and dehydroepiandrosterone sulfate (DHEA-SO4) to cardiovascular risk factors in postmenopausal women." *American Journal of Epidemiology* 1995; 142:925–34.

Barrett-Connor, E., Goodman-Gruen, D., et al. "Dehyroepiandrosterone sulfate does not predict cardiovascular death in postmenopausal women: The Rancho Bernardo Study." *Circulation* 1995; 91:1757–60.

Labbate, L. A., Fava, M., et al. "Physical fitness and perceived stress: relationships with coronary artery disease risk factors." *Psychosomatics* 1995; 36:555–60.

Slowinska-Srzednicka, J., Malczewska, B., et al. "Hyperinsulinaemia and decreased plasma levels of dehydroepiandrosterone sulfate in premenopausal women with coronary heart disease." *Journal of Internal Medicine* 1995; 237:465–72.

Pate, R. R., Pratt, M., et al. "Physical activity and public health: a recommendation from the Centers for Disease Control and Prevention and the American College of Sports Medicine." *Journal of the American Medical Association* 1995; 273:402–7.

Mitchell, L. E., Sprecher, D. L., et al. "Evidence for an association between dehydroepiandrosterone sulfate and nonfatal, premature myocardial infarction in males." *Circulation* 1994; 89:89–93.

Denti, L., Pasolini, G., et al. "Correlation between plasma lipoprotine Lp(a) and sex hormone concentrations: A cross-sectional study in healthy males." *Hormone Metabolism and Research* 1994; 26:602–8.

Haffner, S. M., Mykkanen, L., et al. "Relationship of sex hormones to lipids and lipoproteins in nondiabetic men." *Journal of Clinical Endocrinology and Metabolism* 1993; 77:1610–15.

Stampher, M. J., Hennekens, C. H., et al. "Vitamin E consumption and the risk of coronary disease in women." *New England Journal of Medicine* 1993; 328:1444–9.

Rimm, E. B., Stampher, M. J., et al. "Vitamin E consumption and the risk of coronary heart disease in men." *The New England Journal of Medicine* 1993; 328:1450–6.

Eich, D. M., Nestler, J. E., et al. "Inhibition of accelerated coronary atherosclerosis with dehydroepiandrosterone in the heterotopic rabbit model of cardiac transplantation." *Circulation* 1993; 87:261–69.

Barrett-Connor, E., Khaw, K. T., et al. "A prospective study of dehydroepiandrosterone sulfate, mortality, and cardiovascular disease. *New England Journal of Medicine* 1986; 315:1519–24.

Slowinska-Srzednicka, J., Zgliczynski, S., et al. "Decreased plasma levels of dehydroepiandrosterone sulphate (DHEA-S) in normolipidaemic and hyperlipoproteinaemic young men with coronary artery disease." *Journal of International Medicine* 1991; 230:551–53.

References

Nestler, J. E., Clore, J. N., et al. "Dehydroepiandrosterone: The "missing link" between hyperinsulinemia and atherosclerosis?" *The FASEB Journal* 1992; 6:3073–75.

LaCroix, A. Z., Yano, K., et al. "Dehydroepiandrosterone sulfate, incidence of myocardial infarction, and extent of atheroschlerosis in men." *Circulation* 1992; 86:1529–35.

Nestler, J. E., Barlascini, C. O., et al. "Dehydroepiandrosterone reduces serum low density lipoprotein levels and body fat but does not alter insulin sensitivity in normal men." *Journal of Clinical Endocrinology and Metabolism* 1988; 66:57–61.

CHAPTER FIVE

Bortz, W. D., Bortz II, W. M. "Sexuality and aging—usual and successful." *Journal of Gerontology* 1996; 51A:M142–46.

Schiavi, R. C., Segraves, R. T. "The biology of sexual function." *Clinical Sexuality* 1995; 18:7–23.

Morales, A. J., Nolan, J. J., Nelson, J. C., and Yen, S. C. "Effects of replacement dose of dehydroepiandrosterone in men and women of advancing age." *Journal of Clinical Endocrinology and Metabolism* 1994: 78:1360–67.

Belsky, G. "10th national survey: Americans & their money." *Money,* May 1995, 24–5.

CHAPTER SIX

Morales, A. J., Nolan, J. J., Nelson, J. C., and Yen, S. C. "Effects of replacement dose of dehydroepiandrosterone in men and women of advancing age." *Journal of Clinical Endocrinology and Metabolism* 1994; 78:1360–67.

Nasman, B., Olsson, T., et al. "Serum dehydroepidandrosterone sulfate in Alzheimer's disease and in multi-infarct dementia." *Biol Psychiatry* 1991; 30:684–90.

Spath-Schwalbe, E., Dodt, C., et al. "Deydroepiandrosterone sulphate in Alzheimer's disease (letter)." *Lancet* 1990; 335:1412.

Ozasa, H., Kita, M., et al. "Plasma dehydroepiandrosterone-to-cortisol ratios as an indicator of stress in gynecologic patients." *Gynecol Oncol* 1990; 37:178–82.

Medical essay. Supplement to Mayo Clinic Health Letter. "Alzheimer's disease: Living with a 'long goodbye.' " 1996; 1–8.

Parker, L. N., Levin, E. R., Lifrak, E. T. "Evidence for adrenalcortical adaptation to severe illness." *Journal of Clinical Endocrinology and Metabolism* 1985; 60:947–52.

Regelson, W., Kalimi, M. "Dehydroepiandrosterone (DHEA)—the multifunctional steroid: Effects on the CNS, cell proliferation, metabolic and vascular, clinical and other effects. Mechanism of action?" *Annals New York Academy of Sciences* 1994; 719:564–75.

CHAPTER SEVEN

McLachlan, J. A., Serkin, C. D., et al. "Dehydroepiandrosterone modulation of lipopolysaccharide-stimulated monocyte cytotoxicity." *Journal of Immunology* 1996; 156:328–35.

Brind, J. "Spotlight on DHEA: a marker for progression of HIV infection?" (Editorial) *Journal of Laboratory and Clinical Medicine* 1996; 127:522–3.

Beer, N., Jakubowicz, D. J., et al. "Dehydroepiandrosterone reduces plasma plasminogen activator inhibitor type 1 and tissue plasminogen activator antigen in men." *American Journal of the Medical Sciences* 1996; 311:205–10.

Okabe, T., Masafumi, H., et al. "Up-regulation of high-affinity dehydroepiandrosterone binding activity by dehydroepiandrosterone in activated human T lymphocytes." *Journal of Clinical Endocrinology and Metabolism* 1995; 80:2993–96.

Danenberg, H. D., Ben-Yehuda, A., et al. "Dehydroepiandrosterone (DHEA) treatment reverses the impaired immune response of old mice to influenza vaccination and protects from influenza infection." *Vaccination* 1995; 15:1445–48.

Regelson, W., Kalimi, M. "Dehydroepiandrosterone (DHEA)—the multifunctional steroid: Effects on the CNS, cell proliferation, metabolic and vascular, clinical and other effects. Mechanism of action?" *Annals New York Academy of Sciences* 1994; 719:564–75.

Hennebold, J. D., Daynes, R. A. "Regulation of macrophage dehydroepiandrosterone sulfate metabolism by inflammatory cytokines." *Endocrinology* 1994; 135:67–75.

Wisniewski, T. L., Hilton, C. W., et al. "The relationship of serum DHEA-S and cortisol levels to measures of immune function in human immunodeficiency virus-related illness." *American Journal of the Medical Sciences* 1993; 305:79–83.

Garg, M., Bondada, S. Reversal of age-associated decline in immune response to Pnu-imune vaccine by supplementation with the steroid hormone dehydroepiandrosterone. *Infection and Immunity* 1993; 61:2238–41.

Araneo, B. A., Shelby, J., et al. "Administration of dehydroepiandrosterone to burned mice preserves normal immunologic competence." *Archives of Surgery* 1993; 128:318–25.

References

Araneo, B. A., Ryu, S. Y., et al. "Dehydroepiandrosterone reduces progressive dermal ischemia caused by thermal injury." *Journal of Surgical Research* 1995; 59:250–62.

Findling, J. W., Buggy, B. P., et al. "Longitudinal evaluation of adrenocortical function in patients infected with the Human immunodeficiency virus." *Journal of Clinical Endocrinology and Metabolism* 1994; 79:1091–96.

CHAPTER EIGHT

Personal communication. Genelabs Technologies, Inc. 1996.

Genelabs Technologies, Inc. News release: Genelabs initiates second Phase III trial of GL701 (DHEA) in lupus. Redwood City, CA, 1996.

Suzuki, T., Suzuki, N., et al. "Low serum levels of dehydroepiandrosterone may cause deficient IL-2 production by lymphocytes in patients with systemic lupus erythematosus." *Clinical Experimental Immunology* 1995; 99:251–55.

Van Vollenhoven, R. F., Engleman, E. G., McGuire. "An open study of dehydroepiandrosterone in systemic lupus erythematosus." *Arthritis & Rheumatism* 1994; 37:1305–10.

Denburg, S. D., Carbotte, R. M. et al. "Corticosteroids and neuropsychological functioning in patients with systemic lupus erythematosus." *Arthritis & Rheumatism* 1994; 37:1311–20.

CHAPTER NINE

McCormick, D. L., Rao, K. V. N., et al. "Exceptional chemopreventive activity of low-dose dehydropepiandrosterone in the rat mammary gland." *Cancer Research* 1996; 56:1724–26.

Secreto, G., Zumoff, B. "Abnormal production of androgens in women with breast cancer." *Anticancer Research* 1994; 14:2113–18.

Schwartz, A. G., Pashko, L. L. "Cancer prevention with dehydroepiandrosterone and nonandrogenic structural analogs." *Journal of Cellular Biochemistry, Supplement,* 1995; 22:210–17.

Kelloff, G. J., Boone, C. W., et al. "Mechanistic considerations in chemopreventive drug development." *Journal of Cellular Biochemistry,* Supplement 1994; 20:1–24.

National Cancer Institute, Chemoprevention Branch and Agent Development Committee. "Clinical development plan: DHEA analog 8354." *Journal of Cellular Biochemistry,* Supplement 1994; 20:141–46.

Kelloff, G. J., Crowell, J. A., et al. "Strategy and planning for chemopreventive drug development: Clinical Development Plans." *Journal of Cellular Biochemistry,* Supplement 1994; 20:55–299.

Massobrio, M., Migliardi, M., et al. "Steroid gradients across the cancerous breast: An index of altered steroid metabolism in breast cancer?" *Journal of Steroid Biochemistry and Molecular Biology* 1994; 51:175–81.

Inano, H., Ishii-Ohba, H., et al. "Chemoprevention by dietary dehydroepiandrosterone against promotion/progression phase of radiation-induced mammary tumorigenesis in rats." *Journal of Steroid Biochemistry and Molecular Biology* 1995; 54:47–53.

Schwartz, A. G., Pashko, L. L. "Cancer chemoprevention with the adrenocortical steroid dehydroepiandrosterone and structural analogs." *Journal of Cellular Biochemistry* 1993; 17G:73–9.

Regelson, W., Kalimi, M. "Dehydroepiandrosterone

(DHEA)—the multifunctional steroid: Effects on the CNS, cell proliferation, metabolic and vascular, clinical and other effects. Mechanism of action?" *Annals New York Academy of Sciences* 1994; 719:564–75.

Phipps, W. R., Martini, M. C., et al. "Effect of flax seed ingestion on the menstrual cycle." *Journal of Clinical Endocrinology and Metabolism* 1993; 77:1215–19.

CHAPTER TEN

Marin, P., Lonn, L., et al. "Assimilation of triglycerides in subcutaneous and intraabdominal adipose tissues in vivo in men: effects of testosterone." *Journal of Clinical Endocrinology and Metabolism* 1996; 81:1018–22.

Ettinger, B., Friedman, G. D., et al. "Reduced mortality associated with long-term postmenopausal estrogen therapy." *Obstetrics & Gynecology* 1996; 87:6–12.

Davis, S. R., Burger, H. G. "Clinical review 82: Androgens and the postmenopausal woman." *Journal of Clinical Endocrinology and Metabolism* 1996; 81:2759–63.

Rako, S. *The hormone of desire: The truth about sexuality, menopause, and testosterone.* Harmony Books, 1996.

Casson, P. R., Facquin, L. C. "Replacement of dehydroepiandrosterone enhances T-lymphocyte insuline binding in postmenopausal women." *Fertility and Sterility* 1995; 63:1027–31.

Castelo-Branco, C., Martinez de Osaba, M. J., et al. "Circulating hormone levels in menopausal women receiving different hormone replacement therapy regimens." *Journal of Reproductive Medicine* 1995; 40:556–60.

Zumoff, B., Strain, G. W., et al. "Twenty-four-hour mean

plasma testosterone concentration declines with age in normal premenopausal women." *Journal of Clinical Endocrinology and Metabolism* 1995; 80:1429–30.

Casson, P. R., Facquin, L. C., et al. "Replacement of dehyroepiandrosterone enhances T-lymphocyte insulin binding in postmenopausal women." *Fertility and Sterility*, 1995; 63:1027–31.

Touitou, Y. "Effects of ageing on endocrine and neuroendocrine rhythms in humans." *Hormone Research* 1995; 43:12–19.

Morales, A. J., Nolan, J. J., Nelson, J. C., and Yen, S. C. "Effects of replacement dose of dehydroepiandrosterone in men and women of advancing age." *Journal of Clinical Endocrinology and Metabolism* 1994; 78:1360–67.

Miller, K. L. "Alternatives to estrogen for menopausal symptoms." *Clinical Obstetrics and Gynecology* 1992; 35:884–93.

Rotter, J. I., Wong, F. L., et al. "A genetic component to the variation of dehydroepiandrosterone sulfate." *Metabolism* 1985; 34:731–736.

CHAPTER ELEVEN

Balieu, E. E. "Dehydroepiandrosterone (DHEA): A Fountain of Youth?" *Journal of Clinical Endocrinology and Metabolism* 1996; 81:3147–51.

Butler, R. N. "A wake-up call for caution (editorial)." *Geriatrics* 1996; 51:15–16.

Casson, P. R., Hornsby, P. J., Buster, J. E. "DHEA: Panacea or Palaver." Speaker's Abstracts: American Menopause Society, 7th Annual Meeting, Chicago, IL 1996.

Kelloff, G. J., Boone, C. W., et al. "Mechanistic considerations in chemopreventive drug development." *Jour-*

nal of Cellular Biochemistry, Supplement 1994; 20:1–24.

Hornsby, P. J. "Current challenges for DHEA research." *Annals of the New York Academy of Sciences* 1995; 774:xiii–xiv.

Tyler, V. E. *The Honest Herbal: A sensible guide to the use of herbs and related remedies.* Pharmaceutical Products Press, 1993; 3rd edition.

CHAPTER TWELVE

Balieu, E. E. "Dehydroepiandrosterone (DHEA): A Fountain of Youth?" *Journal of Clinical Endocrinology and Metabolism* 1996; 81:3147–51.

Silverman, H. G., Mazzeo, R. S. "Hormonal responses to maximal and submaximal exercise in trained and untrained men of various ages." *Journal of Gerontology* 1996; 51A:B30–37.

U.S. Department of Health and Human Services. Physical activity and health: a report of the Surgeon General. Atlanta: U.S. Department of Health and Human Services, Public Health Service, Centers for Disease Control and Prevention, 1996.

NIH Consensus Development Panel on Physical Activity and Cardiovascular Health. "Physical activity and cardiovascular health." *Journal of the American Medical Association* 1996; 276:241–6.

Newcomb, P. A., Klein, R., et al. "Association of dietary and life-style factors with sex hormones in post-menopausal women." *Epidemiology* 1995; 6:318–21.

Opstad, P. K. "Circadian rhythm of hormones is extinguished during prolonged physical stress, sleep and energy deficiency in young men." *European Journal of Clinical Endocrinology* 1994; 131:56–66.

Field, A. E., Colditz, G. A., et al. "The relation of smoking, age, relative weight, and dietary intake to serum adrenal steroids, sex hormones, and sex hormone-binding globulin in middle-aged men." *Journal of Clinical Endocrinology and Metabolism* 1994; 79:1310–16.

Miller, K. L. "Alternatives to estrogen for menopausal symptoms." *Clinical Obstetrics and Gynecology* 1992; 35:884–93.

Phipps, W. R., Martini, M. C., et al. "Effect of flax seed ingestion on the menstrual cycle." *Journal of Clinical Endocrinology and Metabolism* 1993; 77:1215–19.